BRITISH SHIPBUILDING AND THE STATE SINCE 1918

Few industries attest to the decline of Britain's political and economic power as does British shipbuilding in its near disappearance in the course of the twentieth century. On the eve of the First World War, British shipbuilding produced more than the rest of the world put together. But by the 1980s, the industry which had dominated world markets and underpinned British maritime power accounted for less than one per cent of world output.

Throughout this decline, a remarkable relationship developed between the shipbuilding industry and the Government as both sought to restore the fortunes and dominance of this once great enterprise. This book is the first to provide an industry analysis of this period, based on the full breadth of primary sources available. It blends the records of central Government with those of the Shipbuilding Employers' Federation and the Shipbuilding Conference, as well as records from individual yards, technical societies and the trade press.

Lewis Johnman is Principal Lecturer in History at the University of Westminster and Secretary of the British Commission for Maritime History. **Hugh Murphy** is Senior Caird Research Fellow at the National Maritime Museum and a Researcher at the Centre for Business History in Scotland at the University of Glasgow.

Paperback cover image: Pneumatic riveting on the Tyne *c.*1920s (reproduced by kind permission of the Chief Archivist, Tyne and Wear Archives Services)

EXETER MARITIME STUDIES

General Editors: Michael Duffy and David J. Starkey

British Privateering Enterprise in the Eighteenth Century
by David J. Starkey (1990)

Parameters of British Naval Power, 1650–1850
edited by Michael Duffy (1992)

The Rise of the Devon Seaside Resort, 1750–1900
by John Travis (1993)

Man and the Maritime Environment
edited by Stephen Fisher (1994)

Manila Ransomed: The British Assault on Manila in the Seven Years War
by Nicholas Tracy (1995)

Trawling: The Rise and Fall of the British Trawl Fishery
by Robb Robinson (1996)

Pirates and Privateers: New Perspectives on the War on Trade in the Eighteenth and Nineteenth Centuries
edited by David J. Starkey, E. S. van Eyck van Heslinga and J. A. de Moor (1997)

Cockburn and the British Navy in Transition: Admiral Sir George Cockburn, 1772–1853
by Roger Morriss (1997)

Recreation and the Sea
edited by Stephen Fisher (1997)

Exploiting the Sea: Aspects of Britain's Maritime Economy since 1870
edited by David J. Starkey and Alan G. Jamieson (1998)

Power and Politics at the Seaside: The Development of Devon's Resorts in the Twentieth Century
by Nigel J. Morgan and Annette Pritchard (1999)

Shipping Movements in the Ports of the United Kingdom 1871–1913: A Statistical Profile
edited by David J. Starkey (1999)

Coastal and River Trade in Pre-Industrial England: Bristol and its Region 1680–1730
by David Hussey (2000)

The Glorious First of June 1794: A Naval Battle and its Aftermath
edited by Michael Duffy and Roger Morriss (2001)

The Kingston yard of Lithgows on the Lower Clyde with an oil tanker under construction, *c.* 1964.

BRITISH SHIPBUILDING AND THE STATE SINCE 1918

A POLITICAL ECONOMY OF DECLINE

Lewis Johnman
and
Hugh Murphy

UNIVERSITY
of
EXETER
PRESS

First published in 2002 by
University of Exeter Press
Reed Hall, Streatham Drive
Exeter EX4 4QR
UK
www.ex.ac.uk/uep/

British Library Cataloguing in Publication Data
A catalogue record for this book is available
from the British Library.

Hardback ISBN 0 85989 606 4
Paperback ISBN 978 0 85989 607 8

Typeset in 11½pt Monotype Garamond
by XL Publishing Services, Tiverton

Printed and bound by CPI Group (UK) Ltd, Croydon, CR0 4YY

Contents

Tables

Plates

Acknowledgements

We have incurred a huge range of debts in the researching and writing of this volume. All errors, however, remain our responsibility. We would like to thank the staffs of the following institutions: the Bank of England; the Bank of Scotland; Churchill College, Cambridge; the City of Glasgow Archives; Companies House, London; ING/Barings; the Isle of Wight Council Museum Services; the London School of Economics and Political Science; the National Archives of Scotland; the National Library of Scotland; the National Maritime Museum, Greenwich; the National Museum of Labour History, Manchester; Newcastle City Library, Local Studies Section; the Public Record Office, Kew; Southampton City Archives; Tyne and Wear Archives, Newcastle; the University of Glasgow Business Archives Services; and the Watt Library, Greenock.

At the University of Exeter Press thanks to Simon Baker, Genevieve Davey and Anna Henderson.

On a personal level we would like to thank: Sir Robert Atkinson, Bernie Bryant, Jill and Emma Chesworth, George Gardiner, Henry Gilliat, Dr Hugh Hagan, Kate Hutcheson, Nigel and Theresa Johnston, Rose Meisak, Sarah Millard, Professor Michael Moss, the late Jim Nixon, Dr Irene O'Brien, Portia Ragnauth, Moira and Frank Rankin, Lesley Richmond, Alex Ritchie, Professor Tony Slaven, Roy and Gail Smith and Liza Verity. Two people are owed especial thanks: Gill Webster, who produced a manuscript from chaos and disorder, and Dr David J. Starkey, who read and commented on all of the chapters in draft.

Abbreviations

Text

AOR	Auxiliary Oil Replenishment Ship
BAe	British Aerospace
BP	British Petroleum
BS	British Shipbuilders
BSEA	British Shipbuilding Exports Association
C & AG	Comptroller and Auditor General
CMSR	Controller of Merchant Shipbuilding and Repairs
CPA	Committe of Public Accounts
CSA	Clyde Shipbuilders Association
CSEU	Confederation of Shipbuilding and Engineering Unions
DEA	Department of Economic Affairs
DNC	Director of Naval Construction
DSIR	Department of Scientific and Industrial Research
dt	displacement tonnage
DTI	Department of Trade and Industry
dwt	deadweight tonnage
EC	European Community
EEC	European Economic Community
EU	European Union
GEC	General Electric Company
grt	gross registered tonnage
IMEG	International Management and Engineering Group
IMM	International Maritime Marine
IRC	Industrial Reorganisation Committee
JCSS	Joint Committee of Shipbuilders and Shipowners
LPH	Landing Platform for Helicopters
MAC	merchant aircraft carrier
MFC	Maritime Fruit Carriers
mgrt	million gross registered tons
Mintech	Ministry of Technology

MoD	Ministry of Defence
NAO	National Audit Office
NESL	North East Shipbuilders Limited
NSS	National Shipbuilders Security Limited
PAC	Public Accounts Committee
P&O	Peninsular and Oriental Steam Navigation Company
PSOC	Prinipal Supply Officers Committee
RMSP	Royal Mail Steam Packet Limited
SAC	Shipbuilding Advisory Committee
SC	Shipbuilding Conference
SCC	Shipbuilding Consultative Committee
SDC	Shipyard Development Committee
SEF	Shipbuilding Employers' Federation
SIB	Shipbuilding Industry Board
SIC	Shipbuilding Inquiry Committee
SIF	Shipbuilding Intervention Fund
SMFC	Ship Mortgage Finance Company
SRNA	Shipbuilders and Repairers National Association
UCS	Upper Clyde Shipbuilders
ULCC	ultra large crude carrier
USMC	United States Maritime Commission
VLCC	very large crude carrier
VSEL	Vickers Shipbuilding and Engineering Limited
YARD	Yarrow Admiralty Research Department

Notes

BPP	British Parliamentary Papers
LSE	London School of Economics
NMM	National Maritime Museum
PRO	Public Record Office
SIC	Shipbuilding Inquiry Committee
SMFC	Ship Mortgage Finance Company
TWA	Tyne and Wear Archives
UCL	University of Cambridge Library
UGBAC	University of Glasgow Business Archives Centre

Introduction[1]

It is often contended that the British shipbuilding industry has always been a bastion of *laissez faire* individualism on the part of the owners and employers. This, however, is a myth. From their very emergence as recognisable industries, both shipbuilding and shipping developed hand in glove with the State. Indeed, the State's mercantilist economic policies, enshrined in a series of Navigation Acts dating back to the late fourteenth century, extended a vast swathe of protection to both shipbuilding and shipping. It was the Navigation Act of 1660, however, that facilitated a great extension of the mercantile system by asserting that Britain's coastal, imperial and foreign trades were reserved to ships that were British built, owned and crewed. Such legislation was at least instrumental in facilitating the development of British shipbuilding, shipping and seamanship, and remained central to maritime policy until the middle of the nineteenth century.

It was hardly surprising, therefore, that shipbuilders and shipowners viewed the onset of a campaign for, and passing of, free trade and navigation policies in the 1840s and 1850s with deep alarm. The free traders argued that Britain's maritime industries were already enjoying world supremacy, and therefore the expansion of world trade could only mean further success. After a shaky start, the arguments proved wholly justified and from 1860 the maritime industries of Britain boomed. Year upon year from 1870 until after the First World War, Britain launched around 60 per cent of world output, the proportion rising to an astonishing 80 per cent in 1892–94. Having enjoyed hundreds of years of State-supported growth, British shipbuilding came into its inheritance in the free trade era of 1850–1914. Even then, the State did not stand aloof, as shipping benefited from postal subventions to encourage regular steamship sailings on certain routes, while the Admiralty supported the construction of merchant ships that could serve as fleet auxiliaries. The direct beneficiary of such orders was, of course, shipbuilding. If the shipbuilders had viewed the end of protectionism with apprehension, they now adopted free trade with verve and gusto, becoming in the process some of the clearest personifications of late Victorian individualism.[2]

In the nineteenth century, British shipbuilding underwent, in effect, two industrial revolutions. The first rebutted the challenges from the USA and France in mercantile and naval construction. The second witnessed the transformation of an industry based on wood and sail into an industry focused on iron, later steel, and steam. In the process of this transformation, the geographical axis of the industry swung from the south and west of the country to Clydeside, the north of England and Belfast.

This shift was occasioned by a remarkable series of inventions and innovations, and no less a formidable nexus of inventors and innovators. Bell, Elder, Smith, Randolph, Howden, Laird, Carnegie Kirk, Napier, Tod, MacGregor, Wigham Richardson, Leslie, Palmer and Parsons provide only a short roll-call of the great names of the 'golden age' of British shipbuilding and marine engineering. This technological revolution could not have taken place had it not been for a ready supply of relatively cheap labour, both skilled and unskilled but familiar with metal working and engineering. This was supplied from an abundance of general and marine engineering shops and foundries. Furthermore, the new centres of gravity of British shipbuilding were well endowed by geology. The coals and ores of the north of England and Scotland had spawned a vibrant iron and steel industry, while good railway and other transport hubs were also important. Indeed, coal was of huge importance to the shipbuilding trade as its export and domestic transport generated a vast demand for ships. Thus, by dint of factor endowments, entrepreneurial ability and skilled labour, the shipbuilding industry was revolutionised twice in a few decades.

By the second half of the nineteenth century, Britain did indeed 'rule the waves' in both mercantile and naval terms. Its mercantile marine had more than doubled between 1870 and 1914 from around 8.5 million gross registered tons (mgrt) to almost 19 mgrt, a figure that represented over 40 per cent of the world total. This tonnage was almost the sole preserve of British shipbuilders. Strong and enduring bespoke linkages between shipbuilders and shipowners were established, notable examples being the links forged between the Ropner Steam Ship Company and Gray's at Hartlepool; Cunard and John Brown on the Clyde; Lithgows and Lyle Shipping; Scott's and Holt's Blue Funnel Line and Swire's China Navigation; Swan Hunter and the British India Steam Navigation Company and the Palm Line; Vickers and Ellerman's,

Canadian Pacific and Furness Withy. Likewise, Harland and Wolff developed various strong connections with shipping firms, especially the Royal Mail Steam Packet, Elder Dempster and the Bank Line.

Such bespoke relationships had weaknesses as well as strengths. For example, they functioned almost at the level of personal contacts between company chairmen and certainly discouraged a wider and more aggressive marketing strategy. They also implied, if they did not always exactly mean, that building was entirely to order and yards often had to respond to an extremely varied product mix. Thus, while personal contacts played a major role in the British shipbuilding market, shipyard output was largely shaped by buyers, who not only determined the flow of orders, but also the type and range of ships to be built. In other words, British shipbuilding was geared towards production rather than the market.[3] In boom conditions, this orientation was both beneficial and understandable; in a contracting market, however, it could prove dangerous and possibly fatal.

This essentially 'new' industry was a complex assembly trade which ushered in new work practices, skills and trades. New organisational structures developed as boilershops, engineshops, plating sheds and drawing offices were established at the heart of shipyards. This often represented substantial capital investment in plant and equipment in the form of machine tools, cranes, slipways and buildings, all of which required maintenance. Furthermore, to facilitate work on the ways and fitting-out basins, a whole range of services—pressurised air, hydraulic power, water, steam and latterly electricity—all had to be installed, while spur railways were opened to bring in bulky materials. Steel construction also spawned a new range of trades, with boilermakers, riveters, blacksmiths, red-leaders and plumbers becoming key members of the shipyard workforce.

The work undertaken by each trade was organised on a strictly demarcated system that was usually governed by a regionally negotiated rule book. This was necessary because few shipyard workers were employed by one firm but tended to work the length and breadth of the yards on any given river. Given the 'hire and fire' nature of the industry, driven by the available base-load of work, such agreements were essential and demarcation disputes would later become the stuff of industrial relations legend. Throughout its history, however, British shipbuilding was a labour-intensive rather than a capital-intensive industry. It was an atomistic industry wedded to bespoke construction with a high proportion of skilled labourers, all of whom had served a five-year apprenticeship in their respective trades. Organised primarily in squads, these tradesmen worked on a strictly demarcated basis in accordance with their individual trade specialism. Not only did this give rise to management–labour disputes, but it also prolonged inter-union sectionalism in the industry, a situation that the employers in turn exploited and tolerated. As one commentator has observed, in a heavily casualised industry such as shipbuilding, the majority of the workforce was regarded as a 'variable factor of production to be hired and fired according to the employers' short-term labour requirements'.[4] This was a situation that suited the employers in that it reduced the need for middle managers and left much of the organisation of work to the foremen and the squads themselves. Once again, this was a form of work organisation that suited a rising market but nobody thought to ask whether it would be adequate in a falling or more fiercely competitive market.

The downturn of the 1920s would bring the Government back into the industry, albeit hesitantly, in an attempt to stabilise the market. The shipbuilding industry was, however, highly fragmented in structural terms, and this meant that any assistance tended merely to replicate the order patterns of the past. Given the adverse economic conditions confronting a whole swathe of the British staple trades, there were clear limits as to how far any policy of intervention could be pushed. Furthermore, the industry itself set fairly circumspect limits to the forms of assistance which it would accept. Indeed, many in the industry recognised that a failure to act in terms of self-help would be the catalyst for more thoroughgoing intervention by the State. In this regard, the Bank of England encouraged the industry to reform itself rather than await State action.

As with shipbuilding, so too with coal and steel.[5] Given the vested structures of the industry, however, there were very strict limits on how far Government-inspired restructuring could be pushed against a semi-reluctant, albeit desperate, industry and an equally reluctant banking sector, headed by the Bank of England, which feared creeping nationalisation. In one of a whole range of ironies attaching to the history of the British shipbuilding industry, it is doubtless the case that it was the 'outmoded ' nature of much of the business that enabled it to survive the inter-war years. It was just as well, as in retrospect the rise of the Axis powers set a premium on berth capacity, almost any capacity.

Still, without the vast US Liberty ship programme—ironically to a design produced by Thompson's of Wearside—it is doubtful whether the Battle of the Atlantic could have been won. Even this, however, was a double-edged sword in that new construction methods—notably welding to join plates and flow line production systems—were now added to motor propulsion, to form cumulatively a second industrial revolution in metal technology as applied to shipbuilding. Britain had already transformed itself twice in the wood and sail era and had carried all before it in the transition to iron, steel and steam. Could it do the same again in the mid-twentieth century? The portents certainly seemed set fair at the end of the war. The USA could not be expected to maintain its vast merchant shipbuilding industry, given its cost structure, while the major competitor nations of the inter-war years were either knocked out of serious competition (Japan, Germany, Italy) or lay in reduced circumstances (France, Holland, Belgium). Despite taking over 50 per cent of the available orders in 1948, British shipbuilders seemed almost frightened by their success and fearful that the 1918–21 period was about to repeat itself. Indeed, a defensive, steady-state approach to the market would be the hallmark of British shipbuilders' attitudes. Despite all of the evidence to the contrary—that world trade was in a long period of growth and that entering new routes and trades with new cargoes and ships would be at a premium—British shipbuilders stuck to what they knew best: traditional ships, traditional trades and long-standing customers.

This strategy was in part responsible for bringing shipbuilding back to the attention of the State. Seldom can an industry have been so under the microscope without the formulation of any remedial measures. Throughout the 1940s and 1950s, a veritable blizzard of official, quasi-official and unofficial inquiries examined the industry, but only when it looked as if it was about to disappear altogether in the early 1960s was a credit scheme forthcoming. This served to rally the industry's fortunes. A major formal inquiry was mounted in the mid-1960s and the industry was reorganised as a consequence, with considerable financial support. In another irony, however, this once again took place against a strong economic upswing without any consideration as to what would happen in a downswing. It did not take long for this issue to arise. The impact of the two oil price shocks of the 1970s halted the long period of growth that had marked the international economy since the late 1940s and plunged the sea transport industries into recession. World shipbuilding production received its first serious check since the inter-war years. In Britain, the shipbuilding industry was nationalised in the 1970s and then privatised in the 1980s. Having endured a consistently falling share of world markets throughout the second half of the twentieth century,

Britain ceased to be a volume shipbuilder in the 1980s and all but disappeared from the global shipbuilding map in the 1990s.

1

Sea Change
The Impact of the Great War and Boom to Bust in the 1920s

In the last quarter of the nineteenth century and until 1920 British shipbuilders enjoyed almost unrivalled supremacy in world markets. The British mercantile marine had grown from approximately 8.5 mgrt in 1870 to around 19 mgrt by 1913. This huge expansion was almost solely reserved to British shipbuilders. Massive linkages between shipbuilders and shipowners were established and the industry also expanded its share of foreign orders: over 20 per cent of the industry's output in 1913 was for export. This growth was further reinforced by demand from the naval sector with the private yards supplying nearly 60 per cent of orders between 1890 and 1914. In this period the private yards built some 603 naval ships of over 1.78 million displacement tons (dt). This, in its turn, was supported by foreign naval demand, which glutted the yards. Expanding world trade, the extension—'wider still and wider'—of the Empire and the generalised naval race of pre-1914 all served to fill British shipyards to capacity. By 1920 the industry employed well over 300,000 people directly and, by extended linkages, hundreds of thousands more. The dislocation, however, caused by the First World War, which skewed shipbuilding production to naval production, had a profound impact on domestic and international markets. Shipbuilding and shipping came under increasingly tight control and many industrialists wondered whether the extension of controls presaged nationalisation, whilst policy-makers ruminated as to whether or not some form of control may not have been in the industries' interests. When the boom of 1919–21 turned to bust in the later 1920s, the debate on industry and State relations was sharpened. The search for a solution throughout the 1920s was, however, to establish a trope which would be both long-running and long-winded.

Well before the end of the First World War shipbuilders and shipowners expressed their concerns to Government about current and future policy. Both industries had been subjected to increasing Government control to the extent that, at least in the eyes of their owners, they were all but nationalised. The Peninsular and Oriental Line (P&O) Chairman, Lord Inchcape, supported by a formidable array of colleagues from other shipping companies, provided a foretaste of what was to become a familiar plaint. Writing to the Prime Minister, Lloyd George, Inchcape stated that since the beginning of the war, British shipowners had lost 1,417 ships amounting to 4.5 mgrt. The same owners, however, were precluded from ordering replacement tonnage by a combination of naval building and mercantile building on Government account. To Inchcape the pre-war supremacy of the British mercantile marine was being mortgaged, for even though the shipping companies had considerable investment funds available they were effectively debarred from ordering. In his view, shipowners 'should be permitted to put down new tonnage on their own account, with their regular builders as far as possible'. He further declared that if it was the Government's intention to nationalise, shipowners 'may have the opportunity of considering how they can employ their capital, organisations, and energies in other spheres of industry'. All the War Cabinet would do, by way of reply to this extraordinary threat, was to assure Inchcape that the control of shipping was a war measure and was 'not intended' to presage nationalisation.[1]

Inchcape's views were shared by the Committee on Commercial and Industrial Policy advising the Reconstruction Committee on post-war policy in the shipbuilding and shipping industries. This Committee, comprised of shipbuilders, shipowners and marine engineers, expressed its 'grave anxiety' at the present and future positions of the three industries and particularly the 'prospective position'. In the view of the Committee the neglect of merchant shipbuilding had been 'fatally mistaken' and threatened the loss, 'perhaps for ever, . . . [of] that maritime ascendancy which is the essence of our being'. The Committee recommended a mercantile shipbuilding programme of not less than 3 mgrt per annum—a huge ambition given that the current annual rate was running at approximately 1 mgrt per annum—otherwise, 'the shipping and shipbuilding industries of the country, with all that depends upon them, will be compromised beyond redemption'. The combined results of war losses and depreciation through excessive wear and tear would inevitably lead to the mercantile marine being reduced 'to scrap iron for after-war purposes'. Moreover, the reconstruction period would begin 'with a merchant fleet quite unable to meet the demands of trade'.[2] Again, as with Inchcape, the views of the Committee received something of a blasé brush-off, but evidence was beginning to mount that perhaps the industrialists had a better view of the future than a victory-obsessed Government.

The Government, however, though could scarcely ignore the Interim Report of a Board of Trade appointed Committee considering the position of the shipping and shipbuilding industries after the war. Like many other enquiries and reports on the industries down through the ages this report would be strong, both on reviewing history and on future requirements. Like all the others, however, it would be fatally weak on how its suggestions were to be implemented. In historical terms the importance of shipbuilding and marine engineering industries to the national economy was reviewed. Pre-war, they employed well over 200,000 people, capital invested was over £35 million and annual output exceeded a gross selling value of £50 million. Any enquiry into future prospects therefore had to take into account 'their vital importance to the Empire'. The highly atomised nature of the industry was noted. It built every type of vessel, although often on highly bespoke terms with shipbuilders effectively subsidising shipowners, and, whilst it was well equipped, 'the whole industry is apt to suffer . . . from excessive and unreasonable competition for the limited number of orders in the market'. The Committee took a passing swipe at demarcation—totally ignoring the fact that the system had grown up as a result of the transition from wood to steel and sail to steam and had suited both capital and labour—before going on to claim that there were few industries 'where the predominance of British manufacturers has been more marked'. Despite the fact that British output as a proportion of world output had declined from 82 per cent in 1892–94 to 62 per cent in 1910–14, the output of British yards in 1913 still exceeded the rest of the world put together.[3]

The Committee forecast, accurately as it transpired, that there would be a substantial increase in foreign competition after the war as countries recognised the value of national mercantile marines and shipbuilding industries. Indeed, it went as far as to suggest that with the expansion of capacity abroad, that post-war competition was likely to be 'severe', although the acid test for British shipbuilding would be 'later, when the world's most urgent requirements are satisfied'. Amongst a raft of recommendations the Committee urged the amelioration of demarcation to allow the improved use of mechanical appliances and labour-saving tools. The industry, it was felt, should be released from Government control as early as possible at the end of the war and enjoy full powers to buy materials in any market, take advantage of improved banking and marketing facilities, and enhance its research and development and the training of apprentices. The Committee also advocated the use of covered berths, the formulation of a scheme of contra-cyclical building to stimulate the construction of ships on a falling rather than a rising market, and the greater standardisation of all elements of production. Subsidisation of the industry was explicitly rejected, although the Committee recognised the spread of this phenomenon abroad, but it was admitted that if the practice harmed the share of British output of ships and marine engines 'it would become an urgent question for the consideration of the Government . . . [as to] how to maintain the Shipbuilding and Marine Engineering Industries'.[4]

A foretaste of likely competition was available by 1917. The British Embassy in Tokyo reported that 'the wholesale withdrawal of foreign vessels from most routes in the Orient has been attracting the attention of shipping circles in this country'. In the view of the Embassy, the leading Japanese shipping companies had realised that the opportunity was likely to be 'unique' and the Nippon Yusen Kaisha and Osaka Shosen Kaisha were 'both contemplating great schemes of expansion'. The routes in question were: Vancouver to Hong Kong, Australia to Seattle and Tacoma, Bombay to Liverpool via South Africa, Japan to Singapore and Singapore to Australia. The threat to the Bombay to Liverpool Line was obvious and 'would unquestionably lead to much dissatisfaction on the part of British companies'.[5] As if competition in shipping were not bad enough, the British Embassy in Copenhagen noted the growth of the Danish shipbuilding industry and in particular commented on the fact that Burmeister and Wain were specialising in the production of diesel motor vessels with electrically driven deck machinery.[6] Still, however, little of this discommoded Government. At a conference between ministers and shipbuilders and shipowners in March 1918 the view was expressed that control of shipbuilding was unnecessary and that 'facilities for building had been extended during the war to an extent that would provide for all requirements'.[7] Some officials in the Ministry of Reconstruction were, however, prepared to push what they saw as serious issues. For example, some were prepared to advance the view that the shipbuilding industry was the one important industry which had not responded to the pressures of war, having neither indulged in extensive dilution nor in improvement of output via labour-saving devices. Admiralty control of shipbuilding moreover, had failed to break the cycle of distrust existing between capital and labour. According to the employers in the engineering industry, high wages were justified by increased output because a great deal of repetition had been introduced, driving prices down. In shipbuilding the reverse was true and if this state of affairs continued 'the shipbuilding industry will go'. According to the unions, the employers were mainly to blame in that they had never 'mastered the labour saving appliances that have been adopted so commonly in America'.[8] Under pressure from the industry this was a circle which Government could not square.

The belief of some officials, that measures of control over shipbuilding and shipping should continue, remained. Clement Jones, the Assistant Secretary to the War Cabinet, assumed that shipping would continue to be regulated by the Government in the transition period, and that it was 'not improbable that Government control of Shipping . . . has come to stay'. Strong lobbying by the industry, however, tended to override the views of the officials. As Sir Alan Booth, the Cunard Chairman, had remarked, the shipbuilders were 'opposed to any control over shipbuilding after the War, and representative shipowners and shipbuilders . . . as a body did not think that control was necessary and were opposed to it'. At the same time, Sir Alfred Stanley, a minister at the Board of Trade, had concluded that 'British owners desire no protection from foreign competition'.[9] Thus the mood of the shipbuilders and shipowners at the end of hostilities was buoyant and there was a marked desire to end Government influence in the shipbuilding and shipping industries. Confronted by a united front in the maritime industries, and probably mesmerised by the potential scale of the post-war boom, the Government gave way and shipbuilding and shipping were quickly decontrolled. All of the problems and all of the recommendations as to improvements in the future were quickly and quietly forgotten.

The immediate scale of the post-war boom swept away any considerations of problems within the shipbuilding industry. The industry posted record profits, record output and record employment. Certainly few could have predicted that this point would mark the beginning of a period of initially comparative and ultimately competitive decline, which would see British shipbuilding annihilated as an industry of international and national importance. Despite the dislocation in shipbuilding caused by the Great War, the initial post-war years appeared to confirm that the boom would continue. The dislocation, however, had had deeper and permanent ramifications. Competition, which had been growing before 1914, in both shipbuilding and shipping, developed hugely during the war. Expansion spawned an international shipbuilding industry of immense size and productive capacity, while a combination of overseas shipbuilding and economic nationalism led to stiff competition in mercantile trading. Between 1919 and 1922 over 18 mgrt was launched in the world, of which Britain's share increased from 23 per cent in 1919 to 38 per cent in 1922. Much of this growth was, however, illusory. Driven by wartime and post-war inflation, freights had soared and demand for tonnage exceeded supply. In Britain, capacity had been so expanded that the replacement of ships lost proved a relatively easy task. On the north-east coast, for example, the combined capacity of the Tyne and the Wear in 1918 stood at 1.8 mgrt as against launches of 1 mgrt in 1906. In the early 1920s there were 650 berths for the construction of ocean-going ships compared with 450 in 1914. As one Tyneside shipowner put it in 1920:

> the shipowner had less cause for depression than the shipbuilder. . . the shipbuilding facilities of the world . . . [have] approximately doubled. Supposing they resumed the normal increase of trade, such an increase in shipbuilding facilities was evidently quite unnecessary when once the great shortage of tonnage had been overtaken . . . a large number of shipbuilding yards . . . must cease operations.[10]

It was to be an apposite forecast.

The problem, however, was that trade connections had been eroded, goodwill lost and over vast swathes of the oceans foreign shipping lines had captured routes and trades formerly the preserve of British shipping lines. The decline afflicted British ships engaged in deep sea trading, with the Pacific, for example, becoming virtually the preserve of Japanese and American shipping. In the European trades, notably those with the Baltic, Western Europe and the Western Mediterranean, British shipping faced increasing competition from Norway, Sweden, Finland, the Netherlands, latterly Germany, Spain and Italy. One former staple trade, the export of coal, was almost extinguished by the war. Accordingly, a major element of British tramp shipping, the long-distance coal trade, was transformed from a pre-war strength into a post-war weakness. When general freight rates collapsed in 1920 from the boom conditions of 1919–20, with berths at a premium and owners prepared to pay almost any price to renew their fleets and share in the post-war boom, the whole of the shipbuilding industry's cost structure became untenable. For the remainder of the inter-war period the industry would be marked by feeble levels of demand. Indeed, the international mercantile fleet grew from 62 mgrt in 1923 to barely 70 mgrt in 1938, and shipbuilders experienced unprecedented levels of competition in both national and international terms, a combination which induced a decline in Britain's share of international shipbuilding. Of the other major maritime powers, only Germany, whose position was totally and utterly complicated by the reparations settlement, registered a fall between 1914 and 1938. In the rest of the world tonnage grew prodigiously, as Table 1 shows.

Britain's proportion of world tonnage fell consistently during the inter-war period as the problems of weak demand and fierce competi-

Table 1: World merchant tonnage (mgrt) 1914–1938

Country	1914	1938	Percentage change
United Kingdom	18.9	17.8	–5.8
British Dominions	1.6	3.1	93.7
Denmark	0.7	1.1	52.1
France	1.9	2.9	52.7
Germany	5.1	4.2	–17.6
Greece	0.8	1.8	125.0
Italy	1.4	3.3	135.7
Japan	1.7	5.0	194.1
Netherlands	1.5	2.9	93.3
Norway	1.9	4.6	142.1
Sweden	1.0	1.6	60.0
Spain	0.8	0.9	12.5
USA:			
Sea	2.0	8.9	345.0
Lakes	2.2	2.5	13.6
Others	4.0	6.4	60.0
Total	45.5	67.0	46.0

Source: *Lloyd's Register of Shipping.*

tion confronted British shipbuilding and shipping with sectoral decline in world markets. Given the slow growth of the British fleet, very much the staple of British shipbuilding, decline was probable; faced with subsidies, protection and economic nationalism in foreign markets, decline was assured.[11]

None of these problems, however, were obvious in the post-war boom. On the Clyde, John Brown had orders for five liners, including the 20,000–ton *Franconia.* Fairfield's had a larger order book than at any time in the previous 30 years, including liners for the Canadian Pacific, Anchor Line and Anchor-Donaldson Lines. Beardmore's had, in addition to two refrigerated cargo ships, liner orders from the Anchor Line, Cunard and the Lloyd Sabaudo Line. Brown's gained five more contracts in 1919 which filled all the yard's berths, a situation which, the board noted, was 'a condition of matters which has never existed before in the history of the yard'.[12] Between 1920 and 1922, three Tyne yards, Swan Hunter, Palmer's and Armstrong's, individually launched 90,000 gross registered tons (grt) in a single year, whilst on the Wear 318,000 grt went down the slipways in 1920.[13] At Belfast, the Harland and Wolff Chairman, Lord Pirrie, who had already had some success in proseltysing the wartime standard ship, of which 36 were on the stocks at Belfast, reported that the yard had received 22 orders by the end of April 1919, 14 of which had come from the yard's owners, the Royal Mail Steam Packet Group (RMSP). Such was the scale of demand that Harland and Wolff had set in train bold schemes of expansion at Belfast, Greenock and Govan, and was seeking to expand its diesel engine building capacity. It was, however, the bespoke linkages between RMSP and Harland and Wolff which sustained the order book. By the end of 1919, for example, the company had secured a further sixteen orders, twelve from RMSP and four from RMSP's own parent company, the International Mercantile Marine (IMM).[14]

The boom, it seemed, would go on forever. As the *Lloyd's Register* figures for the first quarter of 1919 revealed, the previous world record of tonnage under construction, 3.5 mgrt set in June 1913, had been totally eclipsed by the 7.75 mgrt under construction in the first quarter of 1919. As the trade press was quick to point out, however, whilst 2.25 mgrt were under construction in the UK, over 5.5 mgrt were under construction abroad with over 4 mgrt building in the USA. The UK figure was some 500,000 grt higher than that for the second quarter of 1914, whilst that abroad was 4 mgrt above the June 1914 figure. As one journal noted, the figures were evidence of 'the enormous expansion in shipbuilding in foreign yards'.[15] Commenting on how long the boom would last, Sir Alfred Yarrow argued that it would continue as long as two conditions held: first, that the world's carrying trade in passengers and cargo required more tonnage to carry it, and second, that the UK could supply this tonnage at least as cheaply as foreign competitors. Indeed, Sir Alfred concluded that the future of the whole shipbuilding industry depended upon comparative costs, shipowners being able to order in the UK, remaining in the vanguard of naval architecture and scientific research, and finally taking note of developments in the USA, Japan and Germany where 'the most up-to-date equipment is to be found in their shipyards and engine works'. UK shipbuilders, therefore, had to be bold enough to 'adopt the most modern machines and where the old ones exist . . . take Lord Fisher's advice and "scrap the lot"'.[16] Sir Alfred, however, could have had no idea just how timely, or untimely, his thoughts were.

The situation changed dramatically in 1920 when freight rates peaked and then fell. Much of the boom had been sustained by the fact that shipowners were prepared to pay almost any price for tonnage in order to participate in the post-war boom. Actual and expected inflation since 1914 had underpinned the boom. All costs, materials and wages had soared and the general level of inflation had become such that the Government had instituted deflationary measures in 1920. Production costs generally had risen by over 400 per cent between 1914 and 1920 and the replacement cost of ships was up by approximately 300 per cent in the same period. The trade press was puzzled as to whether the boom had actually broken or was little more than a blip. In June it was reported that the 'remarkable feature' of the shipbuilding industry was the number of enquiries concerning the cancellation of orders, but in the following month 'British shipbuilding supremacy' was being lauded. Tonnage under construction in the UK at the end of June 1920 exceeded the figure for the whole world in the pre-war record year of 1913. Thus, while the prospects for the immediate future were 'highly satisfactory', the fact remained that work was accumulating in the shipyards 'more rapidly than it is being executed'.[17] Confirmation that the situation was serious came at the Furness Withy annual general meeting where the Chairman, Sir Frederick Lewis, stated that whilst the company still had 26 vessels on order it had placed no new orders in 1920. He also cited the example of a subsidiary company which had cancelled two contracts when it realised that the ships would cost double the price quoted when the contract was signed. As Lewis put it, owners were looking closely at their order books because 'of the uncertainties and the almost incessant increases in cost'.[18] Once again, however, the comment would seem eerily apposite as inflation gave way to deflation and shipowners and shipbuilders, so keen to get rid of Government in 1918, began to plead for Government assistance.

In the boom, economic conditions had favoured the shipbuilders at the expense of the shipowners as a strong seller's market prevailed. Spiralling costs, however, did little for efficiency as no builder would quote on a fixed-price basis and orders were only accepted on a time-and-line or cost-plus basis. The trade journal *Fairplay* produced an annual index of the cost of a 7,500 deadweight tonnage (dwt) tramp steamer, and although this index ignores a huge range of questions underpinning it, it does provide an indication of trend (see Table 2). Three trends are obvious; the huge rise in price between 1914 and 1920, a period of comparative stability between 1924 and 1935, and the rise and fall between 1936 and 1939.

By December 1920, the trade press was noting the collapse of the freight market and the cancellation of contracts. Almost all owners

Table 2: Cost of construction of a 7,500 dwt steamer, 1914–1939

Year	Cost (£)	Year	Cost (£)
1914	54,375	1930	67,750
1920	225,000	1931	66,938
1921	97,500	1932	63,094
1922	67,500	1933	62,344
1923	72,188	1934	63,884
1924	68,000	1935	65,594
1925	60,000	1936	72,000
1926	64,500	1937	100,000
1927	66,000	1938	108,000
1928	65,500	1939	100,000
1929	67,250		

Note: Between 1914 and 1926 the figures given are for December in the respective year, thereafter they are for June.

Source: *Fairplay*, January 1940.

became reluctant to continue with orders which had been placed on a cost-plus basis and for which there was no prospect of remunerative employment in the short term. In December, Lloyd Royal Belge cancelled orders for four 10,000 ton passenger-cargo steamers, two at Denny's and two at Scott's. As the Chairman of Denny's put it in the same month, 'six or eight months ago a ship was not to be had except at a ransom price', but now the fall in prices meant that 'shipowning and other trades would suffer somewhat severely'.[19] At Harland and Wolff, eleven orders were delayed and two cancelled outright. The results were a cut in overtime, lay-offs and bonus and wage reductions. Profits duly fell and Pirrie realised that the tie-in with IMM and RMSP did not provide the shelter from the market place which he originally thought had been guaranteed.[20] By April 1921, *Lloyd's Register* reported that nearly 850,000 grt of orders were either suspended or cancelled. In May, Lloyd Sabaudo suspended work on a liner at Beardmore's, Anchor Line had halted work on two liners at Fairfield's, Anchor-Donaldson cancelled one order outright at Clydebank, whilst Cunard suspended work on two liners, including the *Franconia*. By the middle of 1921 the situation in John Brown's was mirrored around the country: four contracts were suspended, two were proceeding at half-pace, one berth was vacant and only one contract was proceeding normally.[21] On the Tyne in 1923, output fell to its lowest level for 30 years at approximately 140,000 grt. In the same year Armstrong's launched under 19,000 grt and Hawthorn Leslie less than 12,000 grt. Armstrong's had six of its ten berths empty in 1923 and the huge Palmer's yard at Jarrow launched just over 11,000 grt in 1925.[22] Within the space of one year, the seller's market, favouring the shipbuilders, had become a buyer's market, favouring the shipowners.

As prices fell bitter battles were fought between the shipbuilders and the trade unions over wages. The unions had scored a notable success in 1919 when a national strike had secured a shorter working week and the maintenance of the wartime bonus. A rising market bestowed power on the unions, a falling market gave that power to the employers. The national pattern, however, was hardly uniform and the unions at Belfast, after a bitter and prolonged struggle, had to settle for a 47-hour week rather than the 44 hours which the unions had demanded. The unions were forced back, after a three-week strike, in February 1919, on the employers' terms.[23] Sporadic scuffles occurred in all districts with the employers adopting the time-honoured method of the lock-out. The issue at national level was first fought out by May 1921 when the unions accepted reductions of 6s per week for time workers and 15 per cent off time rates. Severe as this was, it did little to convince the shipowners, with the Cunard Chairman, Sir Thomas Royden, commenting that whilst the reduction was welcome it would be 'inadequate to permit of construction being resumed'.[24] By August 1921 the scale of the crisis was becoming apparent. Yarrow's announced its intention to close by the end of the year unless 'conditions enable business to be carried on with some chance of success'. As an illustration of competition in the industry, Yarrow had tendered to a British line for a steamer for the India service at a price of £50,000 with no profit, only to see the order placed in Holland at £6,000 less than its tender.[25] By January 1922, 28 per cent of the nation's berths were occupied with normal building, 16 per cent were occupied with suspended or cancelled work, while 56 per cent were completely idle.[26] This was bad enough, but worse was to follow.

The various naval programmes of the Fisher era had been major factors contributing to the profitability of the shipbuilding industry before and during the war. Much was expected of a post-war naval programme and appeared to be confirmed when John Brown, Fairfield, Beardmore and Swan Hunter won contracts for new battlecruisers from the Admiralty. Although the Government had abandoned the commitment to the 'Two Power Standard'—under which there had been a commitment to maintain the Royal Navy at the level of the two closest competing powers—in 1920, it did remain committed to maintaining the Navy as equal to any other in the world. With both the USA and Japan building modern battle fleets, it was widely expected that relief would come in the form of naval orders and the battlecruiser orders appeared to confirm that such would be the case. In December 1921, however, the British Government signed the Washington Naval Treaty which secured parity for the UK with the USA and superiority over Japan without the need for an expensive series of new naval construction programmes. Disarmament suited the broad policy thrust of the British Government. The vigorous pursuit of a deflationary stance could hardly be pursued without reduction in Government expenditures and reducing armaments expenditure certainly chimed with the wider, war-weary, public mood. Seen to be indulging in a renewed arms race on the model of pre-1914 was a luxury that no British Government could afford. Moreover, the Government had promulgated the 'Ten Year Rule' which decreed that the UK would not face a war with a major power for the next decade. The impact of signing the Washington Treaty was immediate: the four battlecruiser contracts were cancelled, and any prospect of a naval programme with substantial numbers of capital ships was rudely dashed. At the stroke of a pen, therefore, the value of naval shipbuilding in 1922–23 fell from £11,816,000 to £721,000, the result being 'the appalling accentuation of unemployment which has followed the withdrawal of Admiralty orders'. Whilst the policy of defence cuts may well have suited the deflationary stance of the Government, it precipitated rising unemployment in the shipbuilding districts, as a deputation of the Lord Mayors of Newcastle and Barrow and the Lord Provost of Glasgow explained to the Admiralty.[27]

Unemployment among insured workers had been negligible during the post-war boom when over 350,000 people were employed in shipbuilding. But it rocketed in the 1920s, as Table 3 shows. To a workforce inured to the cyclical nature of the industry, unemployment on this scale represented something altogether new. Moreover, with job opportunities elsewhere few and far between, long-term unemployment was the fate of many workers.

Although there were periodic glimmers of trade revival, there were longer-term changes taking place which were disadvantageous to the

Table 3: Numbers employed and unemployed in shipbuilding, 1919–1929

Date	Number of insured workpeople	Number unemployed	Percentage unemployed
1919	266,188	12,478	4.7
1920	311,051	16,951	5.5
1921	338,798	39,712	11.7
1922	358,790	126,280	35.2
1923	358,640	123,024	34.3
1924	272,530	86,966	31.9
1925	255,090	80,939	31.7
1926	241,700	89,624	37.1
1927	224,120	82,757	36.9
1928	208,480	42,966	20.6
1929 (Jan.)	202,430	55,947	27.6
1929 (Oct.)	204,720	82,181	40.1

Source: NMM, Shipbuilding Employers' Federation, Statistics.

shipbuilding industry. With one of Sir Alfred Yarrow's conditions for the continuation of the boom, that of the carriage of goods and passengers, already a busted flush, his other, that the UK should remain at the forefront of technical research and keep a weather eye on developments abroad, would also be shown up as hollow. Whilst the trade press could report on new yard layouts, new building techniques, in particular welding, new propulsion techniques, the motorship, and new ship types, e.g. the tanker, all in their diverse ways appeared to pass British shipbuilding by. For UK shipbuilders, however, the immediate problem was tendering. Faced with fierce competition and scarce orders, with prices and delivery dates now the determining factors, the response of management everywhere was to vary tendering practices according to the urgency with which an order was required. During the boom, the standard tender would include the costs of materials and labour for the hull, machinery and outfitting, plus a proportion of overhead charges and an average percentage, of around 10 per cent of the total cost, for profit. Such a contract, known as cost plus, would normally be paid in instalments by the shipowners, according to work progress. By 1920, however, many owners, including Cunard, had changed to payment by bills and they were increasingly seeking quotes on a fixed-price basis. In such a situation, with fixed-price contracts, extended payment facilities and the use of bills, prices were pared to the bone and there was little that shipbuilders could do to influence the situation. The Clydebank Chairman, Sir Thomas Bell, summarised the response of the shipbuilders: 'Each of the larger firms in turn, when short of work, make a plunge, taking an important contract with practically no charges, and the result of this is that the market price on which owners appear to base estimates of what new construction will cost, is fictitiously low.'[28] It was a view which would epitomise the rest of the inter-war period.

Having been anxious to get the Government out of shipbuilding and shipping in 1918, by 1921 both industries were equally keen to get them back in. Much of the argument revolved around the issue of enabling people abroad to purchase goods which they could pay for neither in cash nor by any other immediate means of payment. It was argued that the ordinary mechanisms of commerce and finance could not (or rather, would not) bear such a strain and that the 'Government . . . may be needed to step in with guarantees against ultimate loss'. Despite the fact that this breached sound business principles, such schemes were only advanced 'with the idea of enabling the time to be bridged until ordinary trading conditions reassert themselves'.[29] This, however, was part and parcel of the Government's wider attempt to deal with the problem of rising unemployment. It was initially proposed in 1921 that the State should assist with works which would otherwise be postponed, and should consider the promotion of capital works that would ultimately benefit the UK through the placing of orders in the British Empire and abroad. This scheme was to be financed by a National Development Loan.[30] Although this was approved by the Committee on Unemployment, the Treasury was resistant to the concept of a National Development Loan and suggested the alternative of guaranteeing the interest and sinking fund on loans with the annual liability not to exceed £1 million.[31] The Cabinet considered the various proposals and agreed to consult with the Bank of England. In mid-October it was decided to allocate £25 million to a scheme for the promotion of capital works, a strategy implemented by the Trade Facilities Act, which became law in November 1921. The Act authorised the Treasury to guarantee the payment of interest and principal on loans for the purpose of carrying out capital undertakings or the purchase of articles, other than war materials, which would promote employment in the UK. The aggregate capital guaranteed was not to exceed £25 million, such guarantees to be made within one year and overseen by an advisory committee.[32]

The full gravity of the situation was, however, masked by the accumulated profits of the war and the subsequent boom. It was further occluded by the impact of the Trade Facilities Act. The shipbuilding industry was an early beneficiary of the Act, with Beardmore's securing a loan to resume work on the Italian liner *Conte Verdi*; work also resumed on the *Franconia* at Clydebank and both Harland and Wolff and Palmer's received funding for expansion schemes.[33] Despite indications of a trade revival in 1922, the Chancellor of the Exchequer, Stanley Baldwin, could still term the prospects in shipbuilding as 'less hopeful than any other'. This state of affairs pertained in Baldwin's view because of the general impression that costs in shipbuilding had not bottomed out and therefore shipowners would not order.[34] The extreme uncertainty of the situation was highlighted in the trade press, where in the same column there appeared the following two statements: 'the most depressing . . . [year] for the shipbuilding industry within memory', while there were signs that 'the worst of the depression is passed and a gradual but healthy improvement is taking place'. Costs were further pushed down by another round of wage cuts in 1922: 16s 6d were taken off plain time rates in three stages which resulted in a strike. With very few yards busy, however, the owners defeated the strike and forced another 10s reduction on the resumption of work.[35] Even so, the quarterly shipbuilding figures for June 1922 showed a continued fall in orders, which made it 'clear that the wages reductions . . . have so far done little more than help owners to proceed with suspended work'.[36] The continuing depression prompted the Government to extend the life of the Trades Facilities Act which had been due to expire on reaching its spending limit of £25 million. The Act was duly extended several times between 1922 and 1925 with the limit eventually reaching £75 million.

The early optimism of 1923 soon gave way to renewed pessimism as the industry became embroiled in further industrial strife. A new overtime and night-shift agreement had been negotiated between the employers and the Federation of Engineering and Shipbuilding Trades which was accepted by all affiliated unions with the exception of the Boilermakers' Society. The Boilermakers struck on the Tyne and the response of the Shipbuilding Employers' Federation (SEF) was to lock out all members of the Society on 30 April. The lock-out lasted for seven months and the industry ground to a halt due to the lack of journeymen ironworkers. The dispute was not resolved until July 1924 when a new national agreement on overtime and night-shift working was signed between the SEF and the Boilermakers' Society.[37] It further delayed contracts on which work was proceeding and had a further deleterious impact on ordering. Tonnage launched fell to 66,000 grt during the third quarter of the year and the total for the year, at 645,000 grt, was the lowest total launched since the war. The dispute also had its impact on unemployment in shipbuilding and ship-repairing industries which stood at over 42 per cent of the insured workforce.[38]

Against this background a further round of arguments took place surrounding the impact of the Trades Facilities Act. In the view of the shipbuilders, the industry was so seriously depressed that the interests of viability of plant and labour were well served by the granting of financial facilities to induce new orders. On the other hand, the views of the Shipowners' Parliamentary Committee and the Chamber of Shipping were that with so much surplus tonnage laid up, and with freights depressed, new tonnage merely exacerbated the problem. Indeed, the President of the Chamber of Shipping, Sir Alan Anderson, went so far as to use his 1924 Presidential address to warn of the general undesirability of shipowners seeking State assistance.[39] The Shipowners' Parliamentary Committee also passed a resolution 'that, in the opinion of this Committee, representative of the whole of the shipping industry of the United Kingdom, it is not desirable that the application of the

Trade Facilities Act to shipbuilding should be continued'. The Committee may well have been representative of UK shipowners but when borrowings under the Trade Facilities Act were examined it was revealed that in 1923 alone ten shipping companies, amongst them famous names such as the Royal Mail Steam Packet Company and the Glen Line, had borrowed over £6.3 million for new ships.[40] Ultimately, the Government promised to monitor new shipbuilding applications closely and pledged that the financial limit would not be exceeded.

Various schemes to aid shipbuilding and shipping were considered in 1924 and 1925, such as subsidising the construction of tramp steamers and breaking up obsolete tonnage. But these were all rejected. A Cabinet sub-committee established to examine the issues of assistance to shipbuilding could find no resolution to the problem. As it noted: 'One of the intentions underlying the Trade Facilities legislation is to increase production. In the shipbuilding industry this, in itself, is neither necessary nor desirable, as the amount of tonnage in the world is greatly in excess of present world requirements.'[41] However, it could not quite bring itself to recommend that the Trade Facilities Act should be scrapped. The problem was thrown into sharp, and very public, relief when Furness Withy announced in March 1925 that it had placed an order for five motorships with the Deutsche Werft yard in Hamburg. As British owners rarely ordered abroad, the shock was palpable, with the company being attacked for its unpatriotic action. The reasons—costs and delivery dates—were obvious enough. The lowest British tender was £1,150,000, with delivery of the first ship in fourteen months, as against the German yard's price of £850,000 with delivery in ten months.[42] The position was grave enough for the SEF to call a joint conference with the shipbuilding trade unions on the problem posed by foreign competition and for the ex-SEF President and Chairman of Lithgows, Colonel James Lithgow, to address the Commercial Committee of the House of Commons.[43]

The Joint Conference was rather long on rhetoric and somewhat short on practical suggestions. The employers reviewed the history of foreign competition and conceded that whilst before the war foreign competition had presented a threat, 'the margin of difference then was sufficiently limited' to prevent 'a successful challenge to our premier position and certainly such a catastrophe as an important British order going abroad'. The margin of difference, however, now favoured continental builders and there was every prospect of more British orders going abroad and fewer foreign orders coming to the UK. The only problem in the view of the employers was that foreign workers were prepared to accept poorer working conditions and a lower standard of living than their British counterparts. There was no consideration of yard layouts, capital equipment, construction techniques, propulsion systems or ship types. In a display of blasé indifference that would recur throughout the century, the employers could assert, with no sense of irony, that 'So far as technical equipment, organising capacity, the science of naval architecture, or the skill and craftsmanship of the individual shipyard worker is concerned, this country has nothing to fear in a comparison with either Germany or any other country.'[44] Lithgow's address to the House of Commons Commercial Committee reprised familiar ground. The problem was that 'our lower hours and higher wages' had burdened UK industry 'with a much greater cost than our world competitors'. Once again no consideration was given to factors other than wages and hours, and shipbuilding remained 'a strong, well-organised and virile industry, with pre-eminent technical skill'.[45] Again, it would form the basis of a 20-year trope from the Lithgow camp.

The report from the Joint Conference was published in October 1925 but went little further than recommending an increase in interchangeability. In the meantime, *Lloyd's Register* reported that the last quarter figures for 1925 revealed a fall of 412,000 grt on the equivalent period in 1924, suspended work had increased by 37,000 grt, and only two districts, Belfast and Liverpool, had shown any improvement. Indeed, the annual figures indicated that UK launches had declined by 355,252 grt in 1925 whilst foreign launches had increased by 300,905 grt. Significantly, the statistics revealed that global tanker construction, which had amounted to 1.5 mgrt in 1914, had reached 5.25 mgrt in 1925, and that motorship building had reached 2.75 mgrt in 1925 whereas only 234,000 grt had been in service in 1914. Amidst indications that 1926 might witness an upswing, orders were gained but profits were negligible. Canadian Pacific, for example, ordered twelve vessels, liners, cargo liners and ferries from John Brown's, Fairfield's, Beardmore's and Denny's. Brown's, however, ran up losses on 8 of 23 contracts taken on between 1922 and 1928, with net losses on all 23 amounting to £211,000. The power of the buyer was well illustrated in that John Brown's complained about the owners' consultants, of 'whose rapacity there appears to be no limits'.[46] Of sixteen orders completed by Fairfield between 1924 and 1929 only four were profitable, with three Canadian Pacific contracts being taken at prices which failed to recover costs.[47] Profits and dividends duly suffered. Few companies were able to maintain dividends in the period 1922–26. While Hawthorn Leslie and Swan Hunter and Wigham Richardson did manage this, most, including Blyth, Cammell Laird, Fairfield, William Gray, Harland and Wolff, Northumberland, Palmer's, Ropner's, Thornycroft, Vickers, J. Samuel White and Workman Clark, did not.[48]

Given that shipbuilding, as already stated, had always been afflicted by severe cyclical fluctuations, albeit usually short-lived, the scale of the downswing now facing the industry was dramatic. As one commentator observed, ups and downs and bankruptcies were common, but:

> what was new in the inter-war period was the persistence of over-supply, the determination of other countries to develop their industries at the expense of that of the UK by subsidies and protection, the failure of the shipyards to maintain a lead in most aspects of design and more important to improve the organization of work in the yards. The consequences for employment and for the future prospects of the industry were severe.[49]

Indeed they were. There was a veritable blizzard of closures in the 1920s. On the Clyde, business ceased at Chalmers and Company in Rutherglen, the Campbeltown Shipbuilding Company on the Mull of Kintyre, Ross and Marshall, Dunlop Bremner and Murdoch and Murray in Port Glasgow, Caird's in Greenock, Fullerton's in Paisley, and Ailsa Shipbuilding in Ayr. On the north-east coast, the following Tyneside yards closed: Eltringham's, Rennoldson's, Hepple's, Wood and Skinner, the Newcastle Shipbuilding Company, Dobson's and the Tyne Iron Shipbuilders. The giant Armstrong-Whitworth combine was the subject of a hugely complex rationalisation scheme which, in essence, saw its various shipbuilding concerns taken over by Vickers, in a new company, Vickers-Armstrong. On the Wear and beyond, firms collapsed in quick order: Harkess, Raylton Dixon, Irvine's, Ropner's, Richardson-Duck, Osbourne, Graham and Company, Sunderland Shipbuilding, and Blumers.[50] In addition to Armstrong's, it was well known that other famous names in the industry—Beardmore's on the Clyde and Palmer's on the Tyne—were in deep trouble.

One very high-profile casualty, the Northumberland Shipbuilding Company, is illustrative of attitudes which prevailed in the post-war boom. This firm, located at Howdon on the Tyne, was comparatively insignificant when the north-east shipowner Sir Christopher Furness sold it in 1918. The company's share value was £500,000 but within the year it had been increased to £7 million, of which £5.5 million was in preference shares. By 1920, Northumberland had control of Fairfield's, Irvine Shipbuilding at West Hartlepool, Doxford's at Sunderland, Workman Clark at Belfast, Monmouth Shipbuilding at Chepstow, and, through Workman Clark, the Lanarkshire Steel Company and its associated collieries of John Watson and Company, also in Lanarkshire. Such horizontal and vertical integration had been a significant feature of the pre-war years with firms such as Vickers, Palmer's, Beardmore's and John Brown's all appearing to justify the strategy. Northumberland, however, was a different type of venture. Those behind the group were the London financiers Sperling and Company, who in turn had borrowed heavily from the merchant banker Kleinwort. The problem was that the new company's growth was based entirely on share exchanges, obtaining advances from subsidiaries and pledging assets which resulted in the creation of a highly geared and under-financed group with more than a hint of financial chicanery attached.[51]

The Workman Clark purchase, for example, was based on the fact that the firm's fixed and liquid assets vastly exceeded the value of its share capital. With Workman Clark's liquid cash reserves totalling some £4 million it appeared that Northumberland wished to use the cash reserves to repay loans to Kleinwort's. Fairfield, too, represented an attractive catch, with no bank borrowings, and cash in hand. At face value both companies benefited from their ownership by Northumberland. In February 1920 a £3 million debenture loan was raised, at 7 per cent interest, with the supposed intention of developing Workman Clark; Fairfield's also appeared to benefit, with four sets of Doxford machinery purchased for an oil tanker and for three barges being built at Chepstow, and orders for four tankers for Globe Shipping. Despite the attempt at vertical integration Northumberland was shocked to find that the Lanarkshire Steel Company had never actually rolled any plate and accordingly the company had to order steel from Dorman Long, Cargo Fleet and South Durham Steel and Iron Company. In the event, too much steel was ordered and some had to be cancelled which cost Fairfield's over £323,000 between 1920 and 1921. Two of the Globe tankers were cancelled in the spring of 1921 and Fairfield's had to settle for payment in six-monthly bills. Problems at Chepstow resulted in Fairfield taking control of the yard for £25,000 and then writing down the book value by £66,349. Finally, the Govan yard had to subscribe £400,000 to a £5 million issue of 6 per cent non-cumulative shares in Northumberland Shipbuilding in October 1921. Despite the fancy

Plate 1: The River Wear. The cramped, narrow and tortuous nature of the river is obvious. In some senses it was amazing that the Wear became a major world shipbuilding centre.

financial footwork the depression was taking its toll and when Workman Clark defaulted on the full interest of the £3 million debenture in 1923 legal proceedings were instigated. Eventually the courts decided that the 1920 debenture prospectus had misrepresented the company's position.

It took, however, until 1926 for a receiver to be appointed, although why remains something of a mystery. As *Fairplay* reported in September 1925, Northumberland's accounts for the three years to date were simply awful. There was a debit balance on the profit and loss account of over £390,000 and although net investments approached nearly £9 million, the assets were less than £280,000 with mortgages and debentures over £2 million. Loans from associated companies, mainly Fairfield, totalled £267,166 and led the report to conclude 'that the value of the shares is negligible at the present time'. The fall-out was grim. The once vibrant Workman Clark was neutered, and although various attempts were made to reconstruct the company it did not survive the inter-war depression. Irvine's closed in 1924 and Fairfield's found itself in such reduced circumstances that it all but closed in the same year. Irvine's re-opened in 1927 and, like Fairfield, barely survived the inter-war depression. The Northumberland yard staggered on until 1930, only to become a victim of the industry's attempt at self-help. Discounting crooked behaviour, the Northumberland Shipbuilding Company saga stands as a testament to the ambitions entertained in the boom years. An industry which tended to generate its capital through current earnings, and which, traditionally, did not draw on outside investors, found that earnings were simply not sufficient to provide a return on existing capital. The débâcle also served to damn mergers as a reaction to the depression and further served to increase the desire to maintain independence in what was already a fiercely independent industrial sector. It illustrated two further unpalatable facts: neither extensive combination nor unfettered competition were likely to get the industry through its difficulties.[52]

Although the Trade Facilities Act had been extended in 1925 and 1926, the scheme expired in March 1927. Over its life the Act had disbursed £73 million in guarantees of which shipping and shipbuilding had consumed nearly £20 million. As Table 4 demonstrates, the assistance provided by the Loans Guarantee (Northern Ireland) Act and the Trades Facilities Act meant that public support for the shipbuilding and shipping industries between 1921 and 1926 amounted to over £31 million. So much for facing up to foreign competition.

In spite of various appeals for a revival of the Trade Facilities Act the

Table 4: Guarantees for shipbuilding given under the Trade Facilities Act, 1921–1926, and the Loans Guarantee (Northern Ireland) Acts

	Number of vessels (000)	Tons Gross	Trade Facilities Act	Loans Guarantee Act	Total
			(Amounts of guaranteed loan, £000s)		
British owners:					
North Atlantic	8	129	2,200	1,900	4,100
South Atlantic	18	253	4,080	3,820	7,900
Other trades	95	607	11,290	4,460	15,750
Total	121	989	17,570	10,180	27,750
Foreign owners:	27	225	2,270	1,610	3,880
Total: British and foreign owners	148	1,214	19,840	11,790	31,630

Source: PRO CAB 27/557, Committee on the British Mercantile Marine, December 1933.

Treasury resolutely set its face against a restoration, commenting with regard to shipbuilding that 'any net gain of employment must have been slight'.[53] By 1928 the trade press had begun to advocate, in terms that chimed with the rest of the heavy industries, that the shipbuilding industry should engage in self-rationalisation. The number of shipyards was clearly disproportionate to mercantile demand and likely future naval demand, at least in peacetime. The press went so far as to suggest that it would be a good thing if some of the yards were to shut down: 'Although this may appear to be a drastic remedy, there seems to be no other way out of the difficulties which beset the paths of shipbuilders as a whole'.[54] This led, in large part, to the formation in 1928 of the shipbuilders' own commercial organisation, the Shipbuilding Conference. Although the shipbuilders had hitherto co-operated on labour matters in the SEF, the formation of the Conference at last recognised that mutual co-operation was infinitely preferable to individual annihilation in the market place. The antecedents to this price-protective organisation were very much a result of the apprehension felt by warship firms, which were already hard hit by the overall lack of naval demand, and more so by increasingly cut-throat competition for the remaining pool of orders.[55]

Before this new organisation could bring forward any constructive proposals a further series of inconclusive meetings took place between Government and the industry. The Lord Privy Seal, J.H. Thomas, met a deputation of the SEF in November 1929 but the meeting did little more than go over old ground. Sir James Lithgow, for example, repeated his complaint that the industry was working on very narrow margins, and that the only way to compete was to restrain costs, with any attempt to increase pay likely to be ruinous. The unions, he observed, had been very obstructive in that they were demanding the same piece-work rates even where expensive modern plant had been installed. Thomas sought some refuge by reference to obsolete plant and inefficient methods only to be rebuffed by the SEF Vice-President, Amos Ayre, who expressed the view that there was a very definite limit to the progress which could be made in such terms. The cost of new capital, machinery, maintenance and running costs had proved so heavy 'that yards so equipped were actually less economical in working than some of the older yards with fewer mechanical appliances'. In general terms, however, the industry's profit rate was so poor that the replacement of obsolete plant was all but impossible.[56] Having been on the receiving end of these complaints, Thomas was probably surprised at the SEF's President's review of 1929 which was headlined 'A Year of Achievement: Steadily Increasing Shipyard Work'.[57] Once again, however, any sign of improvement proved to be little more than a straw in the wind.

Despite the fact that *Shipbuilder and Shipping Record* could hail British yards as 'Still the Greatest Shipbuilders' on the basis that Britain had launched, substantially aided by tanker orders for Norwegian owners financed by builders, 54 per cent of the world's output in 1929, the SEF made the full gravity of the situation clear in a memorandum to Government in the spring of 1930. In the view of the SEF:

> The March quarter has been one of the worst experienced for several years in regard to orders for new ships. Orders for cargo boats which have been so notable a feature of recent months have now practically ceased, and several of the firms engaged on liner work are completing the last vessels under construction and dispensing with large numbers of workpeople, since they have been unable to secure other passenger work or warship work to follow that which is being finished.

Noting that there had been a serious decline in warship work, the memorandum continued: '69 per cent of the shipbuilding firms did not receive a single order during January, 84 per cent did not get an order in February, 80 per cent did not obtain an order in March, and 50 per cent of the firms did not receive a single order throughout the whole of the quarter.' The memorandum went on to note that compared to the previous quarter the tonnage commenced in British yards was 14.5 per cent down, while tonnage launched had contracted by 25 per cent. The tonnage for export had reduced by 5 per cent between the two quarters. British yards were being almost entirely sustained by orders for Norwegian tankers although this had the effect, a bad one in the eyes of the shipbuilders, of increasing the proportion of motor tonnage relative to steam tonnage. Unemployment remained high at 27 per cent of the insured workforce, while 'the prices at which orders are obtainable are stated to remain at unremunerative levels'.[58]

Given the intractability of the problems in the industry one of the first moves of the recently formed Shipbuilding Conference was to form a sub-committee charged with the brief of examining schemes to bring capacity and demand into something approaching equilibrium. Although the Conference was formed in 1928, the advent of a minority Labour Government in 1929 undoubtedly raised the spectre of possible legislation on the industry. The declared aims of the Shipbuilding Conference were to combat unfair contract conditions, control tendering abuses, and research and develop means of assistance. It was instructive that one of its first acts was to establish a sub-committee, under Sir James Lithgow, to examine proposals for self-help. This body began its work in November 1928 when Lithgow considered the issue of 'redundant berths'. By early 1929 the sub-committee had formulated the outline of a redundant berth scheme which gained the support of a general meeting of the Shipbuilding Conference in March 1929. At this meeting, approval was granted for 'the organisation by the committee of a scheme having for its objects the purchase of redundant yards, the sterilisation of such yards, and the formulation of a fund by the inclusion in all tenders of a sum calculated on an approved formula'.[59] In 1930 this was to become the National Shipbuilders' Security Ltd (NSS) (see Chapter Two). In an industry noted for its individualism, such developments verged on the revolutionary. While the formation of an employers' association to battle with the unions was one thing, the establishment of an organisation to reduce capacity was quite another. But if the shipbuilders and Government believed that the situation could not get much worse, they were about to be rudely disabused.

The 1920s had seen a virtual revolution. The heady days of the early 1920s, with its strident individualisation on the part of business leaders, would soon give way to incoherent and disparate appeals for some form of assistance. Some stood out, asserting the verities of the free market. Some obvious changes were, however, in train. The Trade Facilities Acts, for example, stimulated both shipbuilding and shipping but, if the sceptics are to be believed, at the cost of both industries. It was, though, an act of direct intervention in a climate which professed itself as being against any form of State intervention in industry. Many of the former supports too had now gone. The Washington Naval Treaty reduced naval building to almost total insignificance. Although world trade did expand, Britain took little part in this. Between 1919 and 1928, for example, the world merchant fleet grew by over 44 per cent as British trade remained sluggish and old markets remained difficult to re-establish. The result was that British trade collapsed, the base-load of British shipbuilding went with it—and as the world fleet expanded so British shipbuilding came under increasing pressure. The outcome was a severe over-supply of shipbuilding which forced freight rates down and further depressed world markets. In terms of world trade Britain could, in theory, have supplied the whole of the world's demand for ships in the inter-war period. The formation of the Shipbuilding Conference and National Shipbuilders' Security Ltd were attempts at 'self-help', but increasingly shipbuilders and shipowners would look to government for some form of support.

2

The Weight of History

The 1930s

If the shipbuilding industry believed that capacity reduction would bring supply into line with demand, it altogether failed to recognise that international economic conditions could get worse rather than better. The whole structure of the world economy, established by the Versailles Treaty, collapsed in 1931 and recession turned into depression. World growth slowed, as did trade, and the strong protectionist inclinations of the 1920s were confirmed in 1931 and 1932 when the UK abandoned in turn the Gold Standard and free trade. The introduction of the Import Duties Act in 1932 did not, however, extend its comforting blanket to the shipbuilding and shipping industries.

In the context of the inter-war years it seems amazing that so many firms were prepared even to attempt to carry on. In an ironic way, the structure of the industry probably contributed to its survival. The lack of modernisation meant that capital costs could be kept down and the burden of adjustment shifted onto labour. Furthermore the bleak years of the late 1920s and early 1930s followed sustained periods of high profitability which, at least for the prudent firm, implied high levels of liquidity. Even where shipyards took 'loss-making' contracts, as long as such contracts covered prime costs—labour and materials—they were sustainable, at least in the short to medium term, and therefore important in ensuring firm survival. Straightforward profit and loss accounts, therefore, may not tell the full story of why so many British shipbuilding companies survived the inter-war depression. As has been observed, despite 'the severity of trading conditions . . . evident in the record of the contracts . . . it was not coming after a period of pressure on liquid resources'.[1] Whilst this may have seemed to mortgage the long term to the short term in that it had hardly admitted of modernisation, firm survival, to the contemporary, was surely more important than long-term planning for an uncertain future. The measures of self-help, the establishment of NSS to reduce capacity and the setting up of the Shipbuilding Conference to combat unfair contract conditions, control tendering abuses and develop means of assistance, availed the industry little.[2] As the Clyde shipbuilder Sir Maurice Denny put it regarding the Conference, it was 'only a loose price protective association . . . [and it was] axiomatic that a conference which exists simply to ensure a profit price to the least efficient member without reference to the necessities of the client is indefensible: it is economically and perhaps ethically wrong'.[3] Self-help then could only take the industry so far. Whilst the arguments regarding capacity would re-emerge at the end of the decade, the bulk of the industry would continue to look to the Government for support in the 1930s.

Under cover of the amorphous concept of rationalisation, Lithgow's sub-committee of the Shipbuilding Conference had made enough progress in formulating an industry scheme for him to solicit the ongoing help of his influential long-term friend Sir Andrew Duncan. Few people were better placed: Duncan was a Director of the Court of the Bank of England and Industrial Adviser to Montagu Norman, the Bank's Governor and a committed advocate of rationalisation. Duncan also had plenty of experience in the shipbuilding industry. He had been Joint Secretary of the Admiralty Shipbuilding Council and had long been associated with the SEF. In April 1929, at a secret meeting at the Bank of England, Lithgow met Duncan and Norman and suggested that the Bank take a hand in rationalising the shipbuilding industry by underwriting the loan capital to enable the subsequent purchase of excess shipbuilding capacity. Norman then outlined a range of conditions precedent to the Bank's support. Any subsequent scheme had to be favourably regarded by Whitehall, but should not result from political influence, and it should not eliminate healthy competition. Moreover, Bank support should be given only to a workable commercial proposition. There was clearly agreement between the parties that rationalisation was desirable, and that Government should be kept firmly on the sidelines throughout the process, and subsequently.[4]

Pressure for some form of intervention was strengthened by the continuing weakness of world economic conditions. The SEF noted the severity of competition at home and abroad. It pointed out that prices at which orders were being obtained did not cover establishment or other charges, that credit terms often made business prohibitive, and that difficulties were being experienced in obtaining overseas payments.[5] Accordingly, as we have seem, with the aid of the Bank of England's rationalisation 'lifeboat', the Bankers Industrial Development Corporation, the industry inaugurated a measure of self-help by incorporating its own rationalisation vehicle, the NSS, in February 1930. The new company raised a nominal capital of £10,000 and had borrowing powers of up to £1 million to be repaid by a 1 per cent levy on the sales of vessels built by the participating members of the Shipbuilding Conference. This was designed 'to assist the shipbuilding industry by the purchase of redundant and/or obsolete shipyards and dismantling and disposal of their contents and the re-sale of the sites under restrictions against further use for shipbuilding'.[6] Whatever the objections advanced against assisting shipping and shipbuilding through the Trade Facilities Act, there could be little doubt that conditions in both industries were grim.

Whilst the NSS prospectus recognised how poor the situation was, it still could not resist being bullish with regard to British shipbuilding. The industry, it was claimed, still held world pre-eminence and shipbuilders had dealt with the issues of reorganisation and modernisation, new technology, new designs and new propulsion systems. It had thus 'been able to meet foreign competition more effectively'.[7] Given that NSS issued this statement in March 1930 the view of the SEF that the March quarter had revealed a serious decline in orders should have given pause for thought. As to foreign orders and foreign competition the SEF reported the views of its member firms that 'more difficulty is said to be experienced in obtaining foreign orders; foreign competition is described as being rather more severe than in recent quarters and is reported to be influencing prices adversely'.[8] Unemployment, on the SEFs own forecasts, was expected to remain at around 30 per cent for the industry for the remainder of the year. Despite the ebullience of the trade press and NSS in 1930, few believed that the situation would improve dramatically but probably even fewer would have thought that things would get spectacularly worse.

The signing of the London Naval Treaty in 1930 extended for a further five years the ban on capital ship construction and imposed other limitations on the size of warships which meant that there could be little relief to shipbuilding from any naval programme.[9] All of the large naval builders were in various degrees of difficulties. Armstrong-Whitworth had been rationalised as part of an amalgamation with Vickers, Palmer's at Jarrow was consistently loss-making and John Brown's lay virtually empty. It was against this background that NSS claimed its first scalp, the Dalmuir shipyard of William Beardmore. Dalmuir had been specifically laid out to engage in naval work and between 1906 and 1919 Beardmore launched four battleships, seven cruisers, twenty-one destroyers, thirteen submarines, six hospital ships and a seaplane carrier. Between 1920 and 1930, however, naval output amounted to one cruiser and two submarines. While there could be no more eloquent testimony to the impact of the Washington Treaty, the London Naval Treaty promised more of the same. Despite the fact that the yard had managed to build eight liners, the volume of work was never enough and losses had mounted remorselessly. By 1929, the company's bankers would extend no more credit and only the Bank of England stood between it and insolvency. With Dalmuir threatening to bring the whole Beardmore group down, the Bank of England provided NSS with £300,000 to purchase and dismantle the yard.[10] The implications were clear: if one of the most modern yards in the country, with an eye to new markets, products and technology, could not survive, then who could?[11]

With Dalmuir sterilised, Sir James Lithgow used the occasion of his presidential address to the Institution of Engineers and Shipbuilders in Scotland to attack the whole thrust of Government policy since the war. What he termed the 'make-shift efforts' to support industry through what was thought to be a temporary depression had to be abandoned:

> Such methods had been the staple of our industrial policy for many years past . . . They had been amplified by each succeeding Government, with the approval of public opinion, with no more definite idea than that something would turn up . . . We would, therefore, be wise to recognise that our diagnosis of the trouble having been inaccurate, our remedies had been equally inappropriate and, therefore, harmful.[12]

This was familiar enough but it rather begged the question of what the Government was to do when it was under such pressure for assistance from shipbuilders and shipowners alike? For example, the SEF was again petitioning Government at the end of 1930 as to how bad the situation in shipbuilding really was. The limp pessimism of May 1930 had given way by December to near panic. The SEF reported that orders for the third quarter of 1930 were only one-seventh of those which had pertained in the fourth quarter of 1929. Only one berth in four was occupied and during the first nine months of 1930 no more than 20 per cent of firms in the industry had received an order. Britain's share of tonnage launched had fallen from over 60 per cent pre-war to nearly 40 per cent in 1930, while Britain's share of the world mercantile fleet had fallen from 39 per cent in 1914 to 29 per cent in 1929. Unemployment in the industry had risen from 23 per cent in December 1929 to 40 per cent in October 1930, and on the Clyde and the north-east coast unemployment had already reached 50 per cent.[13]

Some measure of relief was provided, at least to John Brown's, by the order from Cunard for the vessel which would become the *Queen Mary*. Like all large vertically integrated firms, John Brown's not only suffered from the depression in shipbuilding but also in coal, steel and armour plate. Although Clydebank was not the major drain on its parent company's resources, the fact that the yard had only 125 skilled ironworkers on its books was symptomatic of the state of the order book. The yard lay empty after September 1930. Without the advance of £200,000 from Cunard for the contract, Clydebank's overdraft would have been over 50 per cent of its limit and the yard would have faced closure. To say that the *Queen Mary* contract saved the Clydebank yard may be to overstate the situation but it was a close run thing.[14] This also, however, reflected the situation in shipping where Cunard, increasingly threatened by newer, larger and faster German, American and French liners, particularly on the prestigious North Atlantic route, had decided that market leadership could only be regained with a ship of revolutionary design combining the features of great size, speed and power.[15] But even here the Government was to be drawn in. Although Cunard was in a position to prepare a contract for the vessel by May, the outstanding problem remained the arrangement of adequate insurance cover for the ship whilst under construction and in operation. For this, Cunard looked to the Government.

The association between the State and Cunard in the twentieth century went back to 1903. Then, in the face of the ambitions of J.P. Morgan's International Mercantile Marine, the Admiralty had decided to pay Cunard £90,000 a year to maintain a company of ships of approved speeds and specifications (the genesis of *Lusitania* and *Mauretania*), agreement not being rescinded until December 1927.[16] The problem of insurance cover was neither peculiar to Britain nor to the inter-war period. Prior to the First World War, the market had been very apprehensive about the risks involved in underwriting full insurance cover for large passenger liners. The situation was such that Cunard itself had entered into a mutual insurance scheme with other lines, both British and foreign, an arrangement which had not survived the war.[17] By 1930 there had been little material change in the insurance market and accordingly Cunard had approached the Government with the proposal that whatever insurance risks the market proved unable to take, either in construction or operation, they should be absorbed by the Government. As the Cunard Chairman, Sir Percy Bates, saw it, 'these large and costly ships are relatively few in number, and it is not to be wondered at if the marine insurance market should find itself unable to deal with the specialities'.[18] Bates made it clear that: 'In the ability of the Cunard Steamship Ltd., to finance the construction of such a steamer my colleagues and I have every confidence and we do not ask for assistance in this respect. . . Neither do we ask for subsidies under the guise of inflated mail contracts.'[19] At this stage, therefore, Cunard was seeking nothing other than insurance cover for the vessel.

Advice to the Government on the issue was initially tentative. Sir Richard Hopkins, the Second Secretary at the Treasury, advised the Permanent Secretary, Sir Warren Fisher, that the Governor of the Bank of England be approached with the view of placing the whole of the insurance with the market. The Deputy Controller at the Treasury, G.C. Upcott, took the view that if the Government were to take over a percentage of the insurance of the ship, legislation would be required, and perhaps an awkward precedent established. Sir Charles Hipwood, the Second Secretary at the Board of Trade, endorsing the new orthodoxy which had developed since the ending of Trade Facilities Act, wrote that: 'it has been the settled policy of this country not to subsidise shipping in any way and this policy has been supported by the general body of the shipowners'. The Board of Trade, however, took the view that insurance cover did not constitute a subsidy, as such, and that there were no grounds of general policy for refusing to consider the proposals made by Cunard. Accordingly, the Board accepted that it was in the national interest that the ship should be built, and that if the only way that construction could be guaranteed was through the granting of insurance facilities by the Government, then such assistance should be granted.[20] Hopkins and Fisher, in turn, made the same recommendation to the Chancellor, Philip Snowden, and approval was duly given to the company, with the necessary legislation being enacted as the Cunard (Insurance) Agreement Act 1930.[21]

By mid-1930, other shipping considerations had become paramount with the imminent collapse of the Royal Mail Steam Packet Company. Indeed, as early as October 1929, the Furness Withy Chairman, Sir

Frederick Lewis, had told Sir Horace Hamilton, the Permanent Secretary of the Board of Trade, that he deemed the collapse of the RMSP inevitable. In such circumstances, Lewis felt that Cunard should take over RMSP's main North Atlantic carrier, the White Star Line, whilst Furness Withy took over the other assets, chiefly the Shaw Saville Line. Given the implicit assumption in Government circles that future competition on the North Atlantic depended upon some measure of rationalization, the solution advanced by Lewis appeared to carry some weight. The problem however, was that White Star's parent company, the Oceanic Steam Navigation Company Ltd, was under the control of Morgan's giant but ailing combine, the International Mercantile Marine. Given the hostility which Morgan's activities had entailed before the First World War, any rationalisation rather depended on either Cunard or White Star going bankrupt, or at least becoming economically unviable.[22]

Alarmed by the developing position, the Government undertook inquiries into the RMSP and was horrified to find that the mammoth group's assets were not considered to have a market value of more than £16 million. Furthermore, the group had outstanding Trade Facilities loans of £9.75 million which were secured against assets of £14 million. Thus 'nearly the whole of the assets . . . have been pledged against security for Trade Facilities loans'. Hopkins reasoned that if a collapse was imminent, then the Government should ensure 'that powerful British interests are ready to take the matter over'. He also cautioned that any level of Government involvement might cause the crisis to erupt prematurely and the Bank of England was also requested to monitor the situation. In July 1930, when the RMSP voting trustees decided to sell the White Star Line, Sir Frederick Lewis suggested to the Cunard Chairman, Sir Percy Bates, that Furness Withy and Cunard together should take over the line. Despite the fact that Cunard had recently committed itself, with Government assistance, to building the giant liner, the Cunard board took the view that a joint offer could be made; if rejected, Cunard itself should consider the take-over of White Star. As Bates put it, 'at one time I thought we might have to choose between White Star and building a new ship. Today I think we may face both together.' It was to be baseless optimism, but either way it was clear that Bates was hoping to acquire White Star's assets as cheaply as possible, either through Furness Withy withdrawing or by protracting the negotiations which, given the RMSP crisis, would lower the asset price.[23]

White Star may have been a prime candidate for liquidation but the sensitive political position of the company mitigated against any such course of action. The company was overcapitalised by some £4 million and had current liabilities of £9 million, of which £4 million was loans, including £2 million of Trade Facilities loans. Given this, it was clear that neither the British nor the Northern Ireland Governments, as the secured creditors, could contemplate finding £2 million in their 1931 budgets. Furthermore, it was essential for political reasons, particularly the maintenance of employment in Northern Ireland, that work on the White Star liner *Georgic*, then on the stocks at Harland and Wolff in Belfast, be preserved. By August 1930, Bates had decided to place an independent bid for White Star's North Atlantic assets, with assistance from New York bankers, on the premise that a cash offer would be attractive to White Star's voting trustees. In October, Cunard made an offer of £3.25 million, partly in cash and partly on the assumption of liabilities attached to various ships. The offer involved the take-over of the twelve White Star vessels then operating, which had a book value of just over £6 million, the assumption of and the paying-off of Trade Facilities Acts loans on *Britannic* and *Laurentic*, assurance of liquidity on outstanding amounts and a cash offer. The scheme, however, was declared to be dependent upon State support for the *Queen Mary*, although no approach had been made to the Government, and the Government was not aware that any would be. The offer though failed to make any provision for *Georgic* and, not surprisingly, it was rejected.[24]

By December 1930, Cunard had committed itself to the building of *Queen Mary* which effectively tied up much of its liquid resources and made any cash offer for White Star impossible. Throughout 1931 tortuous and labyrinthine negotiations were conducted to prevent the full extent of the RMSP group's problems becoming public. The group's linkages throughout the economy were gigantic. It owned 140 companies, 15 per cent of the British mercantile marine, comprising over 2.6 million gross tons of shipping. In addition to the White Star group, it owned, amongst others, Union Castle, Elder Dempster, Lamport and Holt, and the Coast and Glen lines. It also owned Harland and Wolff, with its yards at Belfast and on the Clyde, and Harland and Wolff in turn owned the Scottish steel producers David Colville and Sons Ltd. The implications for the wider economy, therefore, of allowing the full-scale collapse of the RMSP group were almost too dreadful to contemplate. With both the Government and the Bank of England attempting to mount a series of rescues and reconstructions, they both took the view that the White Star group could not be allowed to fail as receivership would damn the whole Royal Mail consortium, including Harland and Wolff. The consequences of failure were potentially horrendous: the governments of Britain and Northern Ireland would have to accept huge liabilities on Trades Facilities Act loans; unemployment, already high, would soar; and there would be severe repercussions in the shipping trade, with British prestige gravely compromised.[25]

With this in mind, Montagu Norman warned Bates not to expect White Star at a 'knock down' price. He further indicated that Cunard should take a reasonable view of rationalisation on the North Atlantic and that restructuring was essential. Negotiations dragged on through the spring and summer of 1931 but Bates proved difficult to budge. Thoroughly exasperated, the Bank and the Treasury agreed a deal whereby the Bank advanced £250,000 to White Star and Norman duly informed Bates that the Bank and the Government could keep the White Star going indefinitely and that this would be the course adopted if Cunard did not assist in the rationalisation process.[26] With the onset of the international economic depression, the early optimism of Bates that Cunard could afford to build the *Queen Mary* as well as acquire the White Star Line was shown up as hollow. By December 1931, Cunard's trading position had deteriorated markedly and earnings were not sufficient to cover depreciation.[27] On 12 December the finance houses refused to discount any further bills on *Queen Mary*, and work on the hull was immediately suspended and the workforce of 3,000 at John Brown's was laid off. The suspension came at a time when the industry itself felt that it was facing the worst crisis in its history. *Lloyd's List* in its Annual Review of 1931 led its coverage with the headline 'Shipbuilding: A Record Year of Depression', declaring that 'the year 1931 will stand in the annals of shipbuilding as the most tragic in history'.[28] The suspension symbolised the crisis of the industry, as the local MP recalled 'better a thousand times that a great ship had never been began than that it should have stood mocking us all these weary months, dangling hope before hungry eyes and dashing faith to the pit of despair'.[29]

Bates quickly realised that any appeal to Government for assistance on *Queen Mary* which did not acknowledge the White Star problem would fall on deaf ears. The initial approach was mild enough: 'In view of the altered conditions in the world since the Cunard Insurance Act was passed, we desire to ask for the help of H M Government in the early resumption of work'. Bates then went on to propose that Cunard procure all White Star assets on equitable terms in return for Government-guaranteed funds for the continuation of *Queen Mary* and a second giant liner. In support of this, Bates invoked the 'precedent' of the *Mauretania–Lusitania* agreement, which, he claimed, aimed 'to maintain British prestige and that this was as necessary now as it was then'. He also went to some length to stress the scale of competition, claiming that 'competition in the North Atlantic was with Governments and not with private firms'.[30] To the Treasury, however, assistance was 'undesirable' in that it would re-admit 'the principle of Trade Facilities assistance for shipbuilding and other industries'. The Board of Trade in the meantime had told Sir Percy Bates that 'he should approach the question from this point, namely amalgamation with the White Star . . . [and that] instead of making a direct official request for financial assistance, see in the first place what was the best offer . . . [Cunard] could afford to make for the White Star'.[31] Thus Bates had official confirmation that the twin questions of the liner and White Star would be linked. Without rationalisation there would be no assistance.

Negotiations, however, were complicated and protracted with much exasperation all round. Agreement was finally signed on 30 December 1933, with the merger to take place on 1 January 1934. The Government's side of the bargain became enshrined in the North Atlantic Shipping Act of 16 March 1934, although it had to stand on its head to meet criticism of assistance on such a large scale to the North Atlantic trade. The settlement involved the creation of a new company, not the take-over Bates had desired, the Cunard–White Star line with the shares disbursed 62 per cent to Cunard and 38 per cent to White Star. As a *quid pro quo*, the Government agreed to advance £1 million to Cunard, £2 million to the new company in working capital and also agreement to finance a second giant liner, which would eventually become the *Queen Elizabeth*.[32] As a criticism of Government policy, the point was made in the House of Commons 'that this was in fact a reversal of the Government policy in regard to financial assistance for public works'.[33] This was indeed true, but the Treasury brief for the Chancellor of the Exchequer, Neville Chamberlain, was breathtaking in its insouciance, in that it recommended that the defence should be mounted on the basis of the 'major national interest' which the North Atlantic trade constituted, and that 'proof of this was in the assistance afforded to the Cunard company in 1903'.[34] Thus, having disavowed the 1903 precedent when opposed to further assistance, the Treasury, now that it was advocating assistance, was able to find a remarkable degree of comfort in its existence.

The Government decision was warmly welcomed in public. *The Times*, never exactly slow to criticise subsidies, commented that 'even those who are most critical of the big ships policy will welcome the encouragement given to the hard pressed shipbuilding industry'. Remarkably, it quoted verbatim the view of the Engineering and Shipbuilding Trades Federation, which avowed that the stoppage at Clydebank 'had affected 105,000 workers, that the weekly amount of unemployment insurance paid out amounted to £14,000 and that more than £100,000 had been paid to discharged workers in this way'.[35] Chamberlain, however, consistently denied that unemployment had in any way influenced the Government's thinking, once terming the impact of the resumption of work 'a trifle'. Before the issue was finally settled Sir Percy Bates caused further grave offence, in a long line of such, when he attempted to deny Lord Essendon, the recently ennobled Sir Frederick Lewis, a seat on the Board of the newly reconstituted company. This drew from Sir Warren Fisher the rebuke that: 'had I thought there was any possibility that he would not participate in the guidance of the proposed merger's destinies, my attitude toward the proposition, and, for what it might be worth, my advice to the HMG would have been very different'. As he noted to Chamberlain, 'Sir P. Bates is after a dictatorship for the Cunard by himself', to which Chamberlain replied that 'P.B.'s behaviour is quite intolerable'.[36] Norman also weighed in, declaring that:

> The most important event of the week has been the launching of 534 now called the 'Queen Mary': and your memory will tell you plainly enough that the finishing of this boat was a put-up job to help the Cunard and to give work on the Clyde, the ship being admittedly speculative, expensive and at this stage unwarranted. But such is the force of propaganda! And so easily are we beguiled that we now believe that the new ship is the one thing needed to make this country first and foremost on the Atlantic and that there never was a greater or more far sighted shipowner than Bates. Bates, if you please, who having busted the Cunard but for the help of the Treasury is now already anxious for a second ship under the same conditions.[37]

All in all, it was a remarkable situation. In the space of 20 years the maritime industries had been against Government intervention, for intervention, ambivalent about intervention and pro-intervention. This situation exactly mirrored the commitment of Government, but nowhere did the theoretical questions posed by Sir James Lithgow get a clear answer. But, in the face of the scale of the RMSP group collapse, the various schemes of reconstruction and rescue mounted by the Government staved off a major crisis for the British economy. Government and the maritime industries, it seemed, were inextricably linked.

Whilst the negotiations regarding RMSP went on, the condition of the shipbuilding industry continued to deteriorate. Reviewing the *Lloyd's Register* returns for 1930, the *Financial Times* reported that, with regard to orders, 'the falling off observed in the previous year has continued uninterruptedly'. Commenting that the depression had broadened and deepened leading to nearly 10.5 mgrt of shipping being laid up in the world, the *Financial Times*'s view was that 'the shipping industry in all its branches is at present suffering from the most severe depression within living memory'.[38] Reviewing 'British Sea Power, 1900–1931', the President of the Institution of Naval Architects, Sir Archibald Hurd, revealed thoroughly defensive attitudes which would, in the wider context, serve to inhibit modernisation. Commenting on the large amount of tonnage laid up, Hurd observed that there had only been a small increase in British tonnage since 1913, and that 'if foreign countries had shown the same restraint over the last fifteen years as British owners have done', the world tonnage in existence would have been 13.5 mgrt less. The depression, therefore, was all the fault of the foreigners. More worryingly, when it came to tankers, the building of which had sustained the industry in 1929–30, Hurd suggested that 'we . . . [should] ignore the tanker tonnage, which is a new development unrelated to general international trade'.[39] This was an attitude of mind which would persist even when the transport of oil was one of the main components of international trade. As even the trade press commented, it was 'this tanker tonnage which has kept the shipbuilding industry going for so long', although it was quickly concluded that the greatest activity, mainly powered by diesel engines, was in Denmark, Germany, Sweden and Spain.[40]

The scale of the depression in 1931 reopened the debate on Government intervention. Sir Frederick Lewis was an early advocate, stating that whilst he remained a free-trader by conviction there was no point flying in the face of international economic reality, and that 'some of the chief exponents of free trade in the past have gradually been forced to realise that this country must have the means of bargaining'.[41] The entirely contrary view, however, was taken by the President of the SEF, Richard Green. He described the year 1931 as 'the worst in living memory', claiming that it was necessary to go back to 1887 to find such a small volume of work in hand. Despite terming the suspension of the work on the new Cunarder 'deplorable', he stated that there should be no state intervention or subsidies. The industry, he declared, had 'always been opposed to State aid or interference of any kind', and self-help through NSS represented the best way forward. Foreign shipping and shipbuilding was being sustained by State subsidies and Governments had either to keep paying losses, 'or let the firms run on their own or finally break down'. Most countries, he believed, were realising the worthlessness of subsidies and neither British shipbuilders, nor other industrialists, would ask for them—a view which totally ignored the fact that Sir Frederick Lewis just had. Despite the fact that the percentage of British shipbuilding relative to the world was the lowest it had been for decades, when conditions improved 'we are ready as an industry to fight to get back to our usual 50 per cent figure'.[42] Indeed, it was around the issue of the suspended Cunarder that much angst coalesced. Under the banner headline, 'Should Britain Retire from the North Atlantic?', *Shipbuilding and Shipping Record* attempted an answer. The North Atlantic was regarded as the principal arena where the major reputations in shipping and shipbuilding were won and lost and the Blue Riband remained a potent symbol of national prestige. The problem was that France, Germany, Italy and the USA were all proceeding with large transatlantic liners 'built with considerable State support'. The policy of waiting for something to turn up (i.e. free trade) was now 'useless', while the issue of re-starting the Cunarder 'is not a political one; it is a national one'.[43]

It may well have been the industry view that 'self-help' through NSS was the best way forward, but this was, at best, a double-edged sword. In its short career between 1930 and 1937, when it was converted into a private company, NSS had bought out 28 yards with an estimated total capacity of 1.3 mgrt. Certainly the roll-call of yards closed looks impressive enough, but the headline figures disguise other factors. Some firms were in liquidation. At least one, Renwick and Dalgliesh on the Tyne, had never opened and some had not launched a ship in years. Palmer's at Jarrow, for example, which NSS acquired in 1934, had not launched a ship since 1932. This case illustrates at least one risky element of NSS operations. As a consequence of closure, unemployment in the town of Jarrow reached 70 per cent. It provoked the Jarrow March and led to a high-profile attack on the activities of NSS by the town's MP, Ellen Wilkinson. In *The Town that was Murdered*, she did not actually mention Sir James Lithgow by name, but nor did she need to, everyone knew who the executioner was.[44] A further anomaly was that some yards were kept on a care and maintenance basis and in other cases redundant facilities were taken over by other concerns, notably some of those of Workman Clark by Harland and Wolff, although Harland's itself was only being sustained by the Northern Ireland Government's Loan Guarantee Scheme. Furthermore, whilst it seemed axiomatic that the least efficient and technically obsolete yards should be the first to be closed by NSS, Beardmore's was probably one of the most modern in the country.

The axe also fell in a geographically disproportionate way, with the north-east coast suffering most in numerical terms. This, however, reflected the fact that the depression in shipping mainly afflicted the tramp trades, the building of tramp ships being a north-east speciality and it was in this area of building that excess capacity was most acute. When berth size and purpose (liner and naval tonnage having a much higher added value and work content) are taken into account, it seems probable that closures were reasonably distributed. Destruction of berths was open to criticism in that it did deny any employment opportunities in an upswing. Nor were the difficulties overlooked by those planning the Second World War when shortage of capacity emerged as a problem. So bad was the problem that some of the sterilised and care and maintenance plant had to be re-opened to serve the emergency. Some shipowners were also critical of the scheme, alleging, quite rightly, that its financial underpinning, the levy, fell on them. The levy raised prices to shipowners, who complained that they were paying to raise prices further against themselves through the elimination of competition.

As the 'worst year in living memory' gave way to 1932, the *Lloyd's Register* figures for the June quarter revealed a further decline. The tonnage under construction, 280,692 grt, with 159,000 grt suspended, was 274,911 grt down on the June 1931 figure and was the lowest comparable figure recorded in 50 years. Britain was building just over 25 per cent of world output. The Government realised that it could hardly effect the rescue of RMSP, rationalisation on the North Atlantic and fund the new Cunarder whilst ignoring the rest of the shipping and shipbuilding industries. Furthermore, the Government had adopted protection as the central plank of its economic policy and some industries, notably steel, benefited from tariffs. Sir John Ellerman, for example, wrote to both the Prime Minister, Ramsay MacDonald, and the Lord President of the Council, Stanley Baldwin, to complain about the anomaly of the situation. Claiming that he was the largest taxpayer in the country, despite the fact that his shipping business had run at a loss for years, he had suffered competition from companies and contracts which the Government had supported through the Trades Facilities Act 'which would never have been placed had commercial interests alone been considered'. Now he was faced with Government guarantees to the Cunard for two transatlantic steamers, and 'this practice of assisting an individual line with public money or guarantee and thereby placing such a line in a much more favourable position than another line not so assisted is wrong in principle'. Worse, if the State had to meet losses 'there is no saying where such a policy will lead the Government, in this and other trades'.[45] The problem for the Government, of course, was that it could not stand aside from the wreck of RMSP, and logic therefore dictated that shipbuilding and shipping deserved wider consideration. As the trade press noted: 'The need for protection of other industries has been acknowledged and met by the introduction of tariffs. No similar contribution to the assistance of shipping has been made by the Government, and the time has arrived when the possibility should be considered from all aspects.'[46] The situation would have to get still worse, however, before Government would act.

Consideration of wider Government assistance for shipping followed the Report of the Special Committee of the Chamber of Shipping on Tramp Shipping, which went to the Board of Trade in 1933. This report concluded that as a consequence of the depression 'British tramp shipping has been brought to the edge of bankruptcy; Banks and shipbuilders are deeply involved in shipping; nearly 23 per cent of British tramp tonnage is laid up, and that which is running is unable to earn depreciation required for the replacement of ships that wear out'. The Special Committee recommended the granting of a temporary subsidy 'to enable British tramp ships to meet the competition of foreign ships most of which enjoy subsidies and/or lower working costs'.[47] This, in turn, led to the establishment of a Cabinet Committee on the British Mercantile Marine, chaired by the President of the Board of Trade, Walter Runciman, but also including Stanley Baldwin and the Chancellor of Exchequer, Neville Chamberlain. When the Cabinet Committee met to consider the Tramp Committee report it reviewed a range of conditions and trades. Runciman took the view that any grant of subsidy risked setting off an international subsidy war and would provoke demands from other domestic industries, notably cotton, for subsidies. There could be no return to the Trades Facilities Act, which, whilst it had assisted shipbuilding and provided employment, 'had proved extremely prejudicial to the shipping industry'. Reviewing the various trades, the

Board of Trade considered the position of tanker tonnage to be 'satisfactory'. On cargo liners, the most important section of the British mercantile marine, the Board summarised the financial position of 30 typical British companies in 1932 and concluded that the percentage of dividends to total capital was 1.88 per cent. Whilst most of the companies were making steady, if somewhat small profits, liner values averaged £12.5 per grt and dividends were being paid, although mostly on the interest earned on investments. By contrast, tramp values in the 1932 averaged £7.8 per grt, and 'the value of tramp tonnage was now so low that it could be purchased at about the scrap price, or very little more', and the reserves of the tramp companies were exhausted.[48] The Committee then edged in two directions: consideration of a temporary subsidy for tramp owners and what the likely parameters and impact of a scrapping system would be. No assistance to tanker companies was considered and the position of the cargo liner companies was viewed as, in a sense, already protected, sheltered as they were by the Conference system.

Before the Government would act, however, further evidence was required. The reasons for the position of British tramps were given as falling freights, rising costs and increased competition. Taking 1923 as an index of 100, freights had fallen to just over 62 in August 1933. Costs, however, were about 60 per cent above those prevailing before the war. There was too much tonnage for the trade to be carried. In particular, the coal trade had suffered: coal exports averaged 65 million tons in the period 1911–13 as against 45 million tons in 1930–32. Also, in 1914 some 44 million tons of shipping burnt coal as fuel, a figure which, despite the huge increase in shipping, had fallen to 36 million tons in 1933. In general, the world's seaborne trade was some 6 per cent below that of 1913, whilst world tonnage had increased by 49 per cent and the carrying capacity was up by 75 per cent. Foreign competition had increased, with the volume of entrances and clearances of Spanish, Greek and Italian ships in the River Plate trade having increased by more than five times since the war. The trade also faced competition from liners, largely due to the fact that with the smaller quantities of cargoes moving, it was in many cases more convenient for traders to ship by liners than to charter a tramp which it would be difficult to fill.[49]

With UK output of ships having fallen to 133,000 grt in 1933, just over 27 per cent of world total and the lowest total built for foreign owners, just over 9 per cent, since 1923, it was also decided to investigate support for shipbuilding abroad. These enquiries concluded that in the USA, between 1915 and 1931, Government expenditure on shipbuilding totalled £720 million and loans and guarantees to June 1932 had reached £30 million at between 2.5 and 3.5 per cent. In France, the Crédit Foncier arrangement made £1 million per annum available for four years for shipbuilding with the Government guaranteeing repayment. In Italy, prior to 1928 a £4 million loan for shipbuilding had been advanced and since 1928 loans of £4.4 million had been made available through the Institute per il Credito Navale. There were also bounties on construction and fitting out—£7.5 million between 1926 and 1938—and a scheme for scrapping 600,000 grt of shipping, with a Government grant of 5s 5d per ton of shipping scrapped. In Japan, facilities were advanced through the Deposits Bureau of the Ministry of Finance and the Industrial Bank of Japan, although the amounts were unknown, but under an Act of 1930 £1.5 million was available for building cargo ships. There was also support for a scheme to scrap 400,000 grt of shipping and build 200,000 grt, and the Japanese Government had taken over ships built in Japanese yards at high prices at a loss of some £3 million. In Germany, loans of £3.5 million had been made available and a further loan of £4 million had been allocated to tide the industry over the crisis of 1931–32, in addition to a scrapping scheme worth £600,000.[50] Despite this plethora of evidence however, it was unlikely that the British Government would give direct assistance to shipbuilding and it remained more likely that any positive impact would apply indirectly through shipping and possibly a scrapping scheme.

The public announcement of the Cunard–White Star deal and the assistance to restart work on the *Queen Mary* undoubtedly sharpened the focus. The Chairman of the Aberdeen and Commonwealth Line, O.S. Thompson, described the public reaction as 'shrieks of ill-informed delight', and argued that British shipowners were perfectly entitled to ask for assistance on Empire routes where their business was threatened by 'excessive' State assistance. On the other hand 'unless he can hold his own against his unsubsidised competitors in those trades where on equal terms he has already a sentimental preference, the British Mercantile Marine had better go into liquidation or reform its methods'.[51] The Chamber of Shipping and Liverpool Steam Ship Owners' Association also weighed in, making a formal request that where any section of the British mercantile marine could demonstrate that a subsidy was necessary and would ensure preservation, the Government 'should favourably consider the granting of such a subsidy, taking care not to prejudice other sections of the British shipping industry'. It was also at this point that schemes for scrapping once again came to prominence. Various schemes had been considered in 1930, 1932 and 1933, but all had been rejected for different reasons. Despite the fact that formidable problems remained in formulating any scheme, there were strong compensating reasons for considering the proposal. New construction would be far more efficient than ships built some ten to twenty years ago, which, it was hoped, would substantially increase the scrapping rate. Cargo ship owners were not in a position to finance new construction. Assistance should increase the efficiency of the British mercantile marine, thereby improving its competitive position, and the construction of new tonnage, which would not otherwise have been built, would greatly help the employment position.[52]

Throughout 1934 the Board of Trade amassed a huge amount of evidence on the relative position of the British mercantile marine. The tramp ship owners, the Chamber of Shipping and the Liverpool Steam Ship Owners' Association were deluged by questionnaires and requests for further information.[53] The weight of this evidence seemed to have crystallised, at least in the mind of the President of the Board of Trade, Walter Runciman. In principle, he was still opposed to subsidy but felt that as a policy option it would have to be left open. The calculations were that subsidies could not be denied to cargo liners, with the expected costs being £2 million to the tramp ship owners and £2.5 million for cargo liners, amounting to £4.5 million in any one year. As practical policy, however, 'subsidies could not be recommended on any basis'. A review of tramp shipping, from almost any perspective, revealed that the sector was in a totally parlous state. Accordingly, Runciman argued that there was only one practical policy which avoided the existing volume of tramp tonnage deflating any revival that might occur in the freight market. The suggestion was for a scrap and build scheme whereby tramp owners would sell for breaking around three vessels in exchange for Government assistance on a five to seven year basis for the construction of one modern ship. The result would be the replacement of some 350 to 400 vessels by some 120 modern ships. This would mean a large reduction in tramp tonnage at a propitious time as prices were cheap at around £10 per ton.[54] What would become the Scrap and Build Scheme had been born.

With the North Atlantic Shipping Act on the statute book by March 1934, resolving other issues in shipping and shipbuilding gained heightened focus. The Committee on the British Mercantile Marine met with the shipowners on 24 April, the main conclusion being the tramp ship owners' view that some form of subsidy was essential as the only quick and least dangerous method of preventing the extinction of British tramp shipping.[55] The Cabinet Committee, a few weeks later, rejected subsidies, given the inherent complications, potential reaction and impact on other domestic industries. The idea, however, of retaining the concept idea of a 'fighting subsidy' remained intact, if for no other reason than to encourage the others. Instead the Government preferred the replacement scheme, although certain classes of ship would have to be excluded: tankers, refrigerated ships, those carrying more than twelve passengers, ships solely engaged in the coastwise trade and cargo liners employed in the short sea routes. Assistance, therefore, would be limited to UK tonnage classified as tramp vessels and general cargo liners. A further condition was that only companies with more than 50 per cent of their capital in British hands would qualify. The scheme envisaged the scrapping of some 3 mgrt for the building of 1 mgrt. It would also be restricted to Britain, because Northern Ireland still had its version of the Trade Facilities Act in force and was continuing to book orders under this 'to the considerable annoyance of the shipbuilders of Great Britain'. In discussion, the Committee's only reservation was that the scheme would not prove attractive enough and, as a consequence, would fall flat. As a final observation, Runciman declared that 'it would be the best thing not to be in too great a hurry in making an announcement'.[56]

Indeed, the Government was in no great hurry. Runciman held two meetings with the Tramp Ship Owners' Committee in May and June 1934, even though the owners had not moderated their demand for a general subsidy. The President informed them that it was 'improbable' that the Government would entertain a proposal for a general subsidy and he believed that until and unless a subsidy was ruled out, the owners would not entertain any alternative measures. Indeed, there was no gainsaying the fact that a general subsidy to shipping would spark other demands: 'for example, the cotton industry would claim that, like shipping, it was of national importance and derelict, and should be given the same favourable treatment'. The problem which his scheme for scrap and build faced was that few tramp owners would consider it as long as they thought a general subsidy possible. The only prospect of any subsidy, however, remained the fighting subsidy, the aim of which was to get rid of foreign subsidies. In this discussion an altogether new element was introduced that would have profound ramifications later: the development of air transport.[57] As Runciman observed, it could be the case that in future a much larger proportion of mails would be carried by air rather than by sea, to the extent that 'in three or four years time light mails would not be carried at all by sea'. But it was only with the help of mail subsidies that regular services to India, Australia and New Zealand had been maintained. Mail subsidies could be removed from the shipping lines to the air carriers, with the result that shipping lines might require subsidies to continue to operate. Rationalisation of shipping, therefore, remained the only practicable course.[58]

The Government finally accepted the need for a defensive subsidy for tramp shipping alone and in turn offered to open negotiations with the other major powers with a view to removing subsidies. No replies were forthcoming. Runciman announced the Government's policy in the House of Commons on 3 July 1934. It took, however, until the spring of 1935 to effect the necessary legislation. The British Shipping (Assistance) Act had two parts. First, the Act gave subsidies of not more than £2 million per annum to tramp vessels which would be digressive as freight rates reached the level of 1929. The scheme was administered by the Tramp Shipping Subsidy Administrative Committee which was also to encourage the reorganisation and rationalisation of the industry. Under Part 2 of the Act, a Scrap and Build Scheme was introduced to run for two years. This made provision for the Treasury to offer loans to owners, on favourable terms, for an undertaking to scrap 2 tons of existing tonnage for every 1 ton built. No consideration was given to assisting passenger and cargo liners on the basis that the Conference arrangements already provided a measure of support.[59] The Act, however, was internally contradictory in that the direct subsidy to sailings mitigated against the scrapping of tonnage. It was the second part of the Act of 1935 which, it was hoped, would impact on shipbuilding but the high expectations were not realised. The Ships Replacement Committee examined 74 applications comprising 95 ships of just under 335,000 grt but only approved 37 applications for 50 ships amounting to 186,000 grt. The total cost was £307 million and advances to shipowners amounted to £3.5 million. This, in and of itself, attests to the tepid response to the scheme as the Treasury had been given leave to advance £10 million, at a rate of 3 per cent or less, repayable over twelve years, up to February 1937. The original expectation had been that 600,000 grt would be built and 1,200,000 grt would be scrapped. Only 97 vessels of 386,625 grt were scrapped.[60]

In many respects the scheme was too little, too late, and the contradictory nature of the two parts of the Act fell victim to its timing. Had the scheme been introduced in 1932–33 it could have taken advantage of low freight rates and depressed building conditions. By 1935–36, however, trade was improving, freight rates were rising, with a substantial increase in the autumn of 1936, and shipbuilding was already beginning to benefit from the rearmament programme. Given, however, that the north-east coast was the main beneficiary of the Scrap and Build Scheme, building 34 of the 50 ships, it did support an area not usually identified with naval building. The tramp shipping subsidy implied that an effort would be made to increase freight rates and increase utilisation. The Scrap and Build Scheme depended entirely on the extent to which owners were prepared to scrap existing tonnage, which, given the provisions of Part 1 of the Act, they were unlikely to do on a rising freight market. The scheme was also very rigid. It initially required that all shipbreaking should be in the UK, thereby ignoring the fact that the cost of scrap in Britain was 50 per cent lower than in the international market. The terms for ship construction were nowhere near as generous as those applied to the *Queen Mary*, 3 per cent hardly comparing with 0.5 per cent under the Bank Rate. A further restriction that broken-up ships and new ships should be like for like simply failed to recognise the changes in international trade. It did, however, produce a number of modern ships, although fewer than expected, and did provide some work for the hard-pressed shipyards, although less than had been expected.[61]

The recovery, however, was fitful. The subsidy element of the 1935 Act was extended in 1936 by the passage of the British Shipping (Continuance of Subsidy) Act.[62] The committee charged with responsibility for administering Part 1 of the Act, the Tramp Shipping Advisory Committee, certainly considered it a success. The institution of a number of minimum rate schemes in 1935 had been effective in countering rate cutting by subsidised foreign vessels, and restored the competitiveness of British ships. Freight rates rose and the 1929 base of 100 was exceeded in 1937 when the index reached 135.7. No subsidies were paid in 1937 and the subsidy provisions of the Act ceased at the end of the year.[63] Given that British shipbuilding output had attained over 1 mgrt by 1938, for the first time since 1930, the sudden collapse in the last two quarters of 1938 was deeply worrying. Indeed, freight rates had begun to fall in 1937. This was reflected in the shipbuilding order book in that less than 250,000 grt was under construction in British yards by July 1938. The situation was such that the Government appointed yet another Committee on the Shipping Industry in January 1939 to consider requests for financial assistance made by owners and proposals for assistance to the shipbuilding industry.[64] Given the increasing prospects of war, much was made of the contribution of the maritime industries to national defence. The positions of the coasting trades and tankers were quickly dismissed with one protected and the other facing no particular difficulty. In terms of shipbuilding, the new President of the Board of Trade, Oliver Stanley, considered it:

> essential for the safety of the country to reserve the capacity of the shipbuilding industry to produce a large volume of merchant shipping without delay should war break out. This capacity will not be maintained if the activity of the industry remains at its present low level, since . . . the necessary shipyard labour will be lost to other industries.

Stanley advanced the view that the tramp shipping subsidy should be reinstated, that assistance for liners was worthy of consideration, and that, on 'the urgent grounds of national security', consideration had to be given to any scheme designed to stimulate the building of ships.[65]

By March 1939, the thinking had coalesced around a subsidy of £3 million per annum for tramp shipping, £2.5 million for tramps and plain cargo liners built in the next year, and the provision of £10 million over three years for the building of cargo ships. In terms of shipbuilding it was considered that, excluding warship production and tankers, the industry could produce around 1 mgrt of merchant shipbuilding. The problem, however, was that there had been a substantial decline in merchant shipbuilding, so much so that 'the rate of building at present has fallen substantially below that level and will fall further unless something is done to check and reverse the process, and especially to retain the supply of shipyard labour'.[66] In essence, the prospective exigencies of war concentrated minds. The problem, however, was that with shipbuilding output having fallen to 800,000 grt in 1938 and falling further in 1939, 'at this level of activity the shipbuilding industry would be incapable of fulfilling the estimated war time requirements'. Discussing the proposals of the President of the Board of Trade in March 1939, the Chancellor of Exchequer, although recognising the grave international situation, took the view that the subsidy 'was really a subsidy to meet the high prices quoted by shipbuilders'. The Admiralty, however, argued 'that something must be done urgently to get more shipbuilding and that more tonnage was urgently needed'.[67] Although the Bill was promulgated in July 1939, the outbreak of war intervened and the Bill lapsed on

3 October 1939. How the shipbuilding industry would respond to the rigours of war remained in the minds of many a rhetorical question which caused some alarm.

Despite the fact that the Bill lapsed, the Government promulgated the measures in published proposals and a memorandum on the wartime financial arrangements between the Government and British shipowners. Given that as late as March 1939 there was only 500,000 grt of merchant shipping on the stocks, eight shipyards were totally idle and another nine were completing their last order, the announcement could not have come at a more appropriate time. It led to the placing of orders for approximately 150 vessels of nearly 750,000 grt and, as was later noted, 'as a measure for war . . . [it] can hardly be exaggerated . . . without it many months delay in gearing up for the war would have been experienced, and more importantly it checked the dispersal of skilled labour'.[68] Rearmament also played its part in restoring the fortunes, although progress on this front was grindingly slow given the impact of the Washington and London Naval Treaties. As Table 5 indicates, rearmament only began to have a decisive impact from 1934, but even then output in 1938 was lower than in 1934.

In policy terms, the Government, through the Committee of Imperial Defence, had established a Principal Supply Officers Committee (PSOC) in May 1924, which was reconstituted in 1927 to 'direct peace-

Table 5: Warship construction in private shipyards and Royal Dockyards, 1930–1938 (dt)

Year	Private yards	Royal Dockyards	Total
1930	20,039	5,833	25,872
1931	4,140	12,590	16,730
1932	21,305	17,150	38,455
1933	875	9,790	10,665
1934	60,960	24,015	84,975
1935	21,450	1,980	23,430
1936	66,309	20,260	86,569
1937	97,649	11,720	109,369
1938	80,417	3,270	83,687

Source: Calculated from L. Jones, *Shipbuilding in Britain: Mainly Between the Two World Wars* (Cardiff, 1957), p. 125.

time investigations in respect of all matters connected with supply in war'.[69] PSOC delegated its functions to two bodies, one of which was the Supply Board, which in turn devolved tasks to seven sub-committees, including a shipbuilding sub-committee, Supply Committee III, which was composed of Admiralty and Board of Trade officials.[70] It was not until 1934, however, that Sir James Lithgow joined a three-man Advisory Panel of Industrialists to assist the PSOC in preparations for rearmament and in investigating, planning and reviewing the future requirements of production and supply in wartime.[71] Although planning for wartime continued, it was only in 1937, through the aegis of Supply Committee III, that a Shipbuilding Consultative Committee (SCC) was established. The SCC was chaired by the Burntisland shipbuilder and Chairman of the Shipbuilding Conference, Sir Amos Ayre, with Lithgow and other leading shipbuilders in attendance.[72]

The international situation continued to deteriorate in the mid-1930s. Little store was set by the Anglo-German Naval Treaty of 1935 and when the Japanese walked out of the second London Naval Conference in January 1936 the attitude of the British Government changed. Although the Government remained committed to observing the existing treaties until 12 December 1936, from 1936–37 onwards rearmament quickened and the Government in 1937 gained parliamentary authority to raise £400 million in defence loans to finance a programme estimated to cost £1,500 million over five years. Despite this largesse, defence spending remained subject to financial constraints up to the outbreak of war. The 1934–35 and 1935–36 new construction programmes amounted to less than 149,000 dwt. The 1936–37 and 1937–38 programmes, however, raised the tonnages by 215,000 and 267,000 dt respectively. In total the naval programmes of 1934–35 to 1938–39 provided for the building of 5 battleships, 6 aircraft carriers, 28 cruisers, 59 destroyers, 24 submarines, 6 sloops and 41 other vessels. All told the programmes committed the Admiralty to a total of 169 ships aggregating over 750,000 dt. As had been predicted, however, the attenuation of the armament industries and the combination of a rising naval and mercantile market led to production bottlenecks with the consequence of rising costs and late deliveries. Tribal class destroyers, for example, were taking as long to build as had the pre-1914 Queen Elizabeth battleships, while the costs of the Fairplay plain-cargo steamer had risen dramatically. The cost of wages and materials appears to have risen by some 18 per cent between 1935 and 1938. At almost every level, there were difficulties in securing materials and components, although the shipyards took the opportunity of restoring both their book values and profitability.[73]

Warship procurement was undertaken under a system of so-called price competitive tendering, with a cost-plus element added later. Although this is difficult to quantify, it is probable that the profit rates accruing to the private firms engaged in naval construction between the 1934–35 and 1938–39 programmes rose remorselessly. Despite the generally futile attempts of the Admiralty to compare costs with Royal Dockyard prices, the whole situation was, as one commentator later noted, virtually meaningless owing to the private yards' collusion on prices.[74] Given that most shipbuilders did not expect renewed profitability to outlast the rearmament period it was not surprising that the shipbuilders took the opportunity to increase their liquidity. As Table 6 demonstrates, the naval proportion of estimated rearmament expenditure from 1934 to 1939 averaged 45 per cent of total expenditure. Despite this almost fourfold increase over the period, naval expenditure over the period had almost halved by 1939.

From 1937 onwards, against a background of capacity and labour restrictions, the SCC had, through various hypotheses, attempted to forecast the ratio of new naval and mercantile construction in a wartime scenario. By 1939 this had resulted in a War Scheme, which was adopted by Supply Committee III, the Supply Board and Cabinet. The shipbuilding industry was divided into three categories: naval yards under Admiralty control in which mercantile work would only be placed under exceptional circumstances; 'mercantile only' yards in which the reverse applied; and yards which would undertake more or less equal amounts

Table 6: Estimated annual expenditure on rearmament, 1934–1939 (£000s)

Year (ending March)	Total	Royal Navy	% of total expenditure
1934	32.2	20.9	56
1935	42.6	24.2	57
1936	60.7	29.6	49
1937	104.2	42.0	40
1938	182.2	63.2	35
1939	273.1	82.9	30

Source: PRO, CAB 102/4.

of both types of work. These categories were devised in an attempt to avoid the perceived mistakes of 1914–18 where naval requirements and mercantile building on the Government account had all but emasculated normal mercantile construction. By the outbreak of war, however, the parameters had been set, but in practice they were subject to change as domestic priorities altered and strategic responses to particular crises unfolded. Nevertheless, on the Great War model, it hardly required an element of genius to determine that inevitable losses to enemy activity, whether U-boat, surface raiders, aircraft, mines or accidents in general, would leave the shipbuilding industry, particularly on the mercantile side, hard-pressed to produce and repair more tonnage than was likely to be sunk. Naval new construction and repair took priority, but any production time lags and changes in strategy in this sphere placed an enormous burden on ship repair and conversion. This burden had, for many years, been borne by the Royal Dockyards, but it could not possibly be met without the aid of private yards, a situation which was both detrimental to, and would inevitably impact upon, planned new building.

The SCC, in the course of planning both naval and mercantile work, had already estimated in February 1939 that the berths allocated to mercantile construction, on the assumption that the berths would be fully manned, would produce 180 vessels of over 1.1 mgrt. Given, however, that this assumption was predicated on ridiculously optimistic manpower calculations, it was nonetheless expected that 1 mgrt per annum would be produced with continuous production, with a similar figure attributed to equivalent warship production. The mercantile figure, however, was subject to a further assumption that standardisation of design would occur to save as much labour as possible. At the outbreak of the war, under the preferred scheme, 620,000 grt of shipping, comprising tankers, colliers and coasters, was under construction, as was 135,000 grt of other types, with another 349,000 grt in hand in the preparatory stage but not yet laid down. It was still expected, given sufficient supplies of material and labour, that merchant shipbuilding would achieve a position corresponding to 1 mgrt per annum by January 1940. Again, however, the assumptions were based on individual yards clearing any naval or merchant work under construction and the concentration on one form of construction to the exclusion of the other. Moreover, output was to a large extent reliant on marine and general engineering firms, steel producers and, in the naval sector, those producing armour plate and gun mountings in particular. Further, it would have been advantageous if the shipyards where contracts were placed had had their own engine works, but this linkage too had been hammered in the inter-war period. Without doubt, the onset of war made the whole concept of surviving the inter-war period that more important. It was hardly exact science but the yards had benefited from the rearmament programmes to the extent that it had restored them to profitability. Still, as in 1914–18, it would remain to be seen how the industry would react to the exigencies of war.

In all, the closures of the 1920s and the activities of NSS in the 1930s did little to improve the competitive position of the industry. On the eve of the Second World War, the industry still had a theoretical capacity of 2.5 mgrt. There were, therefore, more idle berths and manpower in 1939 than there was work available for them. Despite the fact that over 180 berths had been eliminated, this simply confused the issue of capacity with berth numbers. As one historian observed, this:

> was just another instance of the failure to realise that output in the future would depend far more on using improved capital equipment and on an entirely different approach to production problems. The key to this development would be even more that of raising the capital needed to completely re-equip the yards in the face of foreign competition as other shipbuilding countries drew ahead in the post-war period.[75]

It was, however, a vain hope. The 1930s stand as a testament to the atomised and highly individualistic nature of British shipbuilding. Even so, it was largely this structure which allowed much of the industry to survive. With the benefit of hindsight it may well have been the case that the shipyards needed to adopt a more capital intensive approach to production but it was exactly the 'outmoded' system of production which allowed the industry to perform an important role in the process of rearmament.

3

The Challenge of War[1]

Given the capacity reductions of the 1930s, the fact that the industry could rise at all to the demands of 'total war' was little short of amazing. Still, as in the First World War, the shipbuilding industry was heavily skewed to the production of naval craft and had it not been for the American programme of Liberty ships, the Battle of the Atlantic may well have been lost. The Liberty ship programme, however, presented as many challenges to British shipbuilding as it solved other problems. It was a shipbuilding programme based on entirely new methods of production, mainly welding and flow line production. The Admiralty would consistently despair of obtaining enough welded structure and it remained to be seen whether the British shipbuilding industry would rise to the new challenges of distinctive construction methods, new trade routes and new cargoes. The scales of history certainly weighed against this but the challenge remained.

When the Admiralty assumed control over all merchant shipping on 26 August 1939, the situation regarding direct control of the private shipbuilding and repair industries had yet to be resolved. By October, with war already declared, an emergency meeting of the Shipbuilding Conference considered what part its members would play in any wartime organisation of the industry. Accordingly, the Conference informed the Board of Trade that if any organisational body were created 'it would not agree to other than fully qualified shipbuilders and marine engineers being in positions of authority and control'.[2] Thereafter, this imperative continued to drive Conference policy, and lobbying continued, even though a new Ministry of Shipping was created which contained a Directorate of Merchant Shipbuilding, headed by Sir Amos Ayre. Ayre's Directorate was subsequently transferred to Admiralty control on 1 February 1940, by which stage the First Lord, Churchill, had already appointed Sir James Lithgow to the Board of Admiralty as Controller of

Merchant Shipbuilding and Repairs (CMSR). Lithgow characteristically refused to take either a salary or allowances for his efforts.[3] Beforehand, in line with Conference policy, Lithgow had made it clear that there 'would be the maximum possible reliance on the goodwill and help of those actually engaged upon the building and repairing of ships and as little interference as possible with the normal methods of carrying on the industry'.[4] Sir Amos Ayre was duly appointed as Lithgow's deputy, whilst Lawrie Edwards of Middle Docks on Tyneside, whose father, George, had held a similar position in the First World War, was given responsibility for ship repair.[5] These appointments had a dual purpose. While they guaranteed the industry's self-interest at the heart of Government for the duration of the war, they also kept other Government departments, such as the Ministry of Supply, at arms length. Conference policy functioned on the premise that it was better to deal with the devil it knew in the Admiralty, rather than with one it did not.

Lithgow or Ayre could hardly complain, given their major roles in NSS, that a situation pertained where capacity constraints and labour shortages threatened production. Neither, for that matter, could politicians who, when in Government had refused to intervene in the activities of NSS on the basis that it was a commercial undertaking and was therefore not within their ambit.[6] It was doubly ironic, therefore, that Lithgow and Ayre, who were now in day-to-day control of the mercantile side of the industry, had to co-operate with organised labour in the national interest, rather than confront it in their own interests. Moreover, in the mercantile sector, employment in new construction had declined by 26,000 since 1937, and if it had not been for the orders placed before the outbreak of war by the Board of Trade in lieu of a Shipping Assistance Act, then very few workers would have been employed at all.[7]

The situation facing the leaders of the industry was compounded by the reality that merchant shipbuilding, starved of investment throughout the inter-war period, would inevitably suffer in comparison to the better-equipped naval yards. Accordingly, it was crucial to the war effort to allow some of the larger mercantile-only firms to concentrate on the production of ships for which they were most suited in order to obtain the maximum output from the available labour, plant and equipment.[8] To this end, Lithgows on the Clyde, Doxford on the Wear, and William Gray at Hartlepool concentrated mainly on volume tramp ship construction for the duration of the war.[9] Composite naval and mercantile work, with the exception of Fairfield on the Clyde, continued to be undertaken at five of the 'big six' Warship Group firms, with one of the six, Harland and Wolff, the largest shipyard in the UK, maintaining eleven out of nineteen berths for mercantile construction.[10] On the Clyde, only three shipyards, Fairfield, Scott's and Yarrow, concentrated wholly on warship construction.

Under the rearmament programme, the larger Warship Group firms at the outbreak of war were already engaged on the construction of the King George V class battleships, and on Illustrious class aircraft carriers. Similarly, other warship firms concentrated on cruisers, destroyers and submarines.[11] Plainly, other outlets had to be found in which to construct the smaller classes of warships. As Buxton has noted, however, the importance of the convoy system and the provision of suitable escort vessels had already been recognised. Nevertheless, the numbers and capability of the latter had been underestimated.[12] Shipping losses had already begun to bite, with 114 Allied ships sunk by U-boats, only twelve in convoy, from September to December 1939. As Roskill noted, though, in the last two months of the year 'the mine had surpassed the U-boat as the principal cause of our shipping losses'.[13] To take, for example, one class of warship, the destroyer, 20 Hunt class vessels had been ordered from Warship Group firms in March 1939 and a further 36 under the War Emergency Programme of 1940. Capacity constraints inevitably led to orders for the smaller classes of escort vessels, such as corvettes being placed in the mercantile sector. In this regard it was fortunate that the firm of Smith's Dock had developed a whaling vessel which was simpler, cheaper and quicker to build than a sloop, and with later modifications this became the basis for the Flower class corvette programme. Sixty of these vessels were ordered. They were considered appropriate for coastal escort duties, had a top speed of sixteen knots, and were shared among sixteen firms which had little previous experience of warship building.[14] Other vessels, such as the Lithgow 'Y' type tramp steamer, and the Warrior tug of Scott and Sons of Bowling on the Clyde, formed the basis of similar orders to other mercantile firms later in the war.[15]

The overall situation of the mercantile yards did not augur well for a successful long-term prosecution of war. In most cases, plant was old, cranage was inadequate, and working conditions in what was essentially a casualised industry were poor. Similarly, industrial relations, if we can dignify the industry with that term, were particularly bad. Consequently, by the outbreak of war a large number of the industry's workers, particularly those in the more elastic fitting-out trades, had left for more secure occupations elsewhere. This meant that the management and workforce that remained tended to be old and thoroughly imbued with a depression mentality.[16] Moreover, the shipbuilding industry at large, whilst not reaching the absolute ceiling of its productive capacity, still had a sizeable gap to surmount, and was in a much weaker state quantitatively and qualitatively than it had been at the outbreak of the First World War.

The Admiralty and the industry, however, had recognised in the light of the fall of France that a deficit in mercantile construction was likely to occur through increased concentration on escorts vessels to counteract enemy sinkings. Accordingly, it was decided to order additional merchant vessels from North America. In September 1940 the Admiralty Merchant Ship Mission left for America and visited numerous actual and potential new entrant shipbuilding firms. The Mission then quickly decided to entrust the building of these vessels to two new yards under the aegis of Todd Shipbuilding Incorporated. Contract documents were signed by 20 December, and seven berths each were newly laid out in two specially constructed Todd yards at Portland, Maine, and Richmond, California, to construct 30 vessels each by December 1942. By April 1941, the first keel had been laid on the first of the 60 Ocean class freighters ordered, which was based on a Sunderland tramp steamer design by the Wear shipbuilder J.L. Thompson. Contemporaneously, 26 similar vessels were ordered from three Canadian yards.[17] Thereafter, the Thompson design formed the basis of the huge American Liberty ship programme, which at its peak in 1943 resulted in eighteen emergency shipyards completing 1,238 vessels at a rate of almost four per day.[18] Subsequently the last Ocean class vessels left the Richmond and Portland yards in July and November 1942, respectively.[19] In the interim, however, the overseas building programmes in the USA and Canada would take time to come to fruition. Clearly, this placed an enormous burden on the British shipbuilding and repair industry, and on the Royal Dockyards, to construct, convert and repair as much tonnage as possible. Owing to production timelags, and enemy attrition, particularly in the early years of the war, the ability to repair and convert vessels was of crucial importance.

From January 1941 to June 1945 (where figures are more reliable) an average of 800,000 grt of shipping was permanently withdrawn from service solely due to repairs. In all, a huge total of over 180 mgrt of cargo-carrying shipping was restored to service, although many of these repairs were of a routine nature.[20] Table 7 shows that at all times from 1941, more manpower was employed on mercantile repairs than on mercantile construction, with the reverse applying on the naval side where the majority of the workforce was engaged for the duration of the war. Whilst the numbers employed on naval construction exceeded those engaged on naval repairs throughout the war, the overall concentration on repairs had a detrimental effect on mercantile new construction in particular. Despite the experience of senior figures in the industry, and officials in government departments, the extent of the repair function, as Postan later noted, 'was perhaps all the heavier for being somewhat unforeseen, or to be more exact, greater than the planners could

Table 7: Shipyard workers employed in the member firms of the SEF in merchant and naval new construction and repairs, first quarters 1941–1945, and September 1945

Year	Naval construction	Merchant construction	Naval repair	Merchant repair	Total
1941	53,500	30,400	32,100	38,800	154,800
1942	57,200	35,900	28,500	47,100	168,700
1943	58,100	39,700	31,400	45,200	174,400
1944	60,100	35,900	31,300	47,100	174,700
1945	55,000	34,700	31,300	41,300	162,300
Sept.	41,600	42,500	20,100	54,200	158,400

Source: NMM, Shipbuilding Employers' Federation, Statistics.

foresee'.[21] By the end of 1939, 111 ships had suffered accidental damage compared with 20 damaged by enemy action. During 1940, 470 naval vessels were damaged, of which nearly half had occurred as a result of enemy action. Ship repairers had, from the middle of 1939, undertaken conversion of ships for anti-submarine duties such as requisitioned large commercial trawlers.[22] Conversion of much larger ships was, however, an important and time-consuming process, with auxiliary anti-aircraft vessels taking from eight to eleven months to convert, and destroyer depot ships taking up to seventeen months.[23]

If the lessons of the First World War were to be heeded, when the U-boat campaign came near to success, the balance between enemy sinkings, new construction, conversion and repairs had to be somehow maintained. Although it has to be emphasised that the UK still had the world's largest mercantile fleet at the outbreak of the Second World War, some 33 per cent of total tonnage, the fleet had declined by 12 per cent since 1914. Nevertheless, with a shortage of suitable escort and anti-submarine vessels, the balance of UK output was likely to be in deficit as the war progressed.[24] Before the entry of the USA into the war, Britain's maritime industries, in particular, and those of her Commonwealth allies, had an awesome responsibility to produce and engine as much tonnage as possible given production constraints and changing priorities. On the naval side, however, from 1941 onwards the repair function was substantially alleviated by American naval dockyards repairing British warships with an estimated output of the equivalent of two major Royal Dockyards.[25]

Although this was of crucial importance, the volume of repairs had been exacerbated by the need to provide merchant vessels with suitable defensive equipment, and by the damage inflicted by magnetic mines, which meant that ships had to be degaussed.[26] As Inman has noted, post-Dunkirk, 73 destroyers alone were laid up in dock, and by March 1941 the alarming figure of over 2.5 mgrt of mercantile vessels under repair or awaiting repair had been reached.[27] The volume of repair work in general inevitably impacted upon the available workforce engaged on new construction at a critical period in the Battle of the Atlantic. The ship-repairing industry in general, subject as it was to the vagaries of time and tide, was notoriously difficult to police. The labour force was casualised, the numbers of firms were fragmented, and fraudulent practice was not uncommon. In the latter respect, senior members of the ship repair firm of F.H. Porter (and an Admiralty recorder who was convicted of receiving petrol in return for favouring the firm) were charged and convicted of fraudulently claiming excess charges from the Admiralty and the Ministry of War Transport to the amount of £750,000 at Liverpool Assizes in October 1942.[28] Moreover, the ship repairers were, if anything, even more conservative than their shipbuilding brethren as it took until 1943 for them to agree to some form of female dilution to augment the male workforce in the industry, whereas in shipbuilding a similar agreement had been negotiated two years earlier.[29]

Given that the industrial relations systems had developed symbiotically with the development of metal and steam production, it remained to be seen whether the tactics employed by the employers during the inter-war years would allow the maximum possible reliance on goodwill, stressed by Sir James Lithgow, to be achieved. As such, even in wartime, the industry's pre-war problems were more likely to resurface rather than disappear. On the mercantile side of the industry, given an apparent shortage of skilled workers, dilution from within the existing workforce, rather than from without, seemed to be an obvious short-term solution to that sector's problems. Numerous factors weighed against a widespread adoption of dilution, however, particularly in the largely inelastic hull trades. By no means the least of these factors was the legacy of strife and unemployment that had scarred the inter-war period when the employers used the lock-out as a method of enforcing industrial discipline, whilst attempting, unsuccessfully in the case of welding, to de-skill certain occupations. Another major factor, in part a consequence of the first, was the widespread conservatism of management and workers alike that had held back the industry, through lack of vision on one side and the worship of restrictive practices on the other.

Although dilution as a process can be used to introduce manual labour into an industry either to do the work previously done by skilled labour, or to partially replace it, in craft-based industries such as shipbuilding the latter course of introducing an element of controlled dilution, combined with apprentice labour, could strengthen rather than weaken the extant system of work organisation. Indeed, this was the overall pattern of dilution in shipbuilding during the war as there was no question of de-skilling the extant labour force. Underpinning this, Inman noted that there was a high proportion of skilled men in the industry in any event, and that this proportion hardly changed throughout the war.[30] This, the need to retain skilled labour, was a factor which, given the imperative to produce, required the maintenance of the status quo rather than its abandonment, a situation that the Government was unlikely to jeopardise. Dilution was also held back by the lack of standardisation and interchangeability in the industry, which meant that training had to be longer in the myriad of strictly demarcated occupational structures of the shipyards. On standardisation, there was, as a highly placed Government official later commented, 'a curious apprehension' on behalf of shipbuilders, that the adoption of such methods would have an adverse effect on the post-war competitive position of the industry.[31] Moreover, for the majority of the inter-war period the lack of sustainable demand, and the intervention of NSS, had discouraged new firms from entering the industry to enhance competition with the established concerns.

The above is a far from exhaustive list of causal factors, and others real and tangential intervened. The exigencies of war, however, had to a large extent temporarily shifted the balance of power in the shipbuilding industry, as its plethora of trade unions, of which the Boilermakers' Society was the most powerful, were now in a far better bargaining position than had hitherto been the case. This altered set of circumstances allowed the unions to insist first on the reinstatement of their unemployed members, and then on the transfer of former shipyard workers from other industries before dilution from within the existing workforce was considered.[32] On transfers, the Boilermakers went so far as to insist that skilled men in less essential industries, employed or unemployed, should be recruited before any general recourse was made to diluted labour.[33] This position was buttressed by Ministry of Labour officials who, to mid-1942, were reluctant, in the interests of industrial harmony, to enforce transfers of labour.[34] Although a national agreement on dilution was signed by the Confederation of Shipbuilding and Engineering Unions (CSEU) in May 1940, agreements with individual unions as to the amount and extent of dilution had still to be negotiated separately. It was not until March 1941, however, that the Ministry of Labour finally resorted to registering skilled labour that had left the shipbuilding industry during the past fifteen years in order to redirect them back into the shipyards at a later date.[35]

Only when this avenue was all but exhausted was an increase in female dilution considered. Womanpower, however, was used sparingly in shipbuilding in comparison to other industries.[36] Nevertheless, if the attitudes of Clyde shipbuilders were symptomatic of the industry as a whole and there is no reason to suspect that they were not, women were not welcome in the shipyards. Women would require 'segregation' within four walls, a separate entrance, and any attempt to introduce them into the shops and berths would serve 'no useful purpose'. Furthermore, even if women were admitted, 'any gain in productivity from their introduction would be negated by a corresponding loss of output from the men employed'.[37] Only from 1942 onwards, despite a national agreement on female dilution dating from July 1941 that discriminated against female workers already employed in the industry, were women employed in greater numbers than before.[38] The majority of women were, however, confined to the shops and sheds away from the berths, although the national picture was hardly uniform. For example, in the east of Scotland district more women were employed in one Government-controlled Royal Dockyard than in five noted private shipbuilding firms.[39]

The return, however, of 15,000 former male employees to the

Plate 2: New technology: women welders at Lithgows during the Second World War.

industry from the middle of 1941 guaranteed, rather than altered, the extant system of work organisation by preserving the nature of the generally uninterchangeable shipyard trades.[40] This remained the case even though an element of interchangeability was inherent in the dilution agreements that were signed. As Inman noted, the introduction of new technology did little to change the situation, as new machinery altered the quantity of labour required rather than its type.[41] The Government, albeit belatedly, did recognise trade union concerns over the extent of dilution, by enacting the Restoration of Pre-War Trade Practices Act in 1942, which promised a return to the *status quo ante* after the cessation of hostilities. Suspicions persisted that dilution would eventually lead to de-skilling of certain occupations, and a diminution of trade union influence. Indeed, such concerns hardened towards the end of the war, even though the passing of the Act meant that the established workforce had in the short term little to fear. Nonetheless, trade union suspicions were in part confirmed by the various extensions of the legislation long after the war had ended.[42] Self-interest predominated and provided the motor for the continued opposition to dilution from the established workforce. Even in 1942, fifteen firms on Clydeside had no dilution whatsoever in the hull trades.[43] That this state of affairs had arisen was hardly aided by manpower planning difficulties and differences of opinion which had arisen between Admiralty and Ministry of Labour officials. As Judges noted, 'planning was on a highly arbitrary model, there was little of the consciously articulated Grand Plan about it'. In his view, compulsion of the workforce was evidently necessary, although, paradoxically, this would require more planning rather than less if female workers were to be placed at the point of greatest need.[44]

A no less intractable situation co-existed, however, in the field of industrial relations, which continued to be standard for the duration of the war. Shipyard managers who had hitherto exercised their rights to hire and fire at will were now constrained from doing so by Ministry of Labour officials under emergency legislation.[45] Skilled labour, owing to scarcity of supply, and in the light of the nominally Socialist-leaning Labour Party's participation in the coalition Government, was in a stronger position. This was a situation that some shipbuilders, particularly on Clydeside, were loath to accept. Indeed, one shipbuilder went as far as to subject Mass Observation staff to 'a two-hour tirade against these animals'. Apparently, this was a far from isolated opinion as it was reported that several of the most important employers interviewed displayed 'an almost pathological hatred of their workmen'.[46] Strikes, even though they had been declared illegal under emergency legislation, occurred throughout the war, and were undertaken mainly in pursuit of higher wages.

On 28 February 1941 a major series of strikes, which subsequently involved over 6,000 general and marine engineering apprentices and which spread to the shipyards, began amongst engineering apprentices at Kilmarnock, Ayrshire.[47] Thereafter, due to well-organised apprentice leaders, the strike soon gripped the west of Scotland and spread across the border to three districts in England. The scale of the apprentices' actions was confirmation of long-held grievances, as four years earlier a similar strike had begun on Clydeside.[48] Parker noted that a number of factors had provoked the apprentices' actions: the fact that diluted labour received, after training, higher wages than fourth- and fifth-year apprentices, the tendency of employers to use apprentices as a source of cheap labour, and 'impatience with the dilatoriness and the negative results of constitutional procedure'.[49] The most radical aspect of this behaviour, bearing in mind that participation could result in imprisonment, was the apprentices' desire that their representatives should separately be allowed to participate in negotiations with the employers. Another prime factor, taking the latter demand into consideration, was union intransigence in not fully supporting their demands.

The scale of the apprentices' industrial action in the west of Scotland was considerable, with 6,662 apprentices on strike in 83 firms, mainly on and around Clydeside.[50] Subsequently, a Ministry of Labour sponsored Court of Inquiry into the dispute was on the whole sympathetic to the apprentices' demands. An Interim Report of 19 March was agreed, which facilitated a return to work pending further negotiations, and by 21 March the matter seemed to have been resolved. Agreement on a new scale of wages for apprentices, which would take effect on the basis of a percentage, according to age, of a time-served journeyman's wage, seemed to have been reached. The strikes continued, however, on Clydeside, Manchester, Barrow and Rochdale, as the latter had remained out on strike and brought the rest out within a week. This clearly exasperated the Minister of Labour and National Service, Ernest Bevin, who apparently decided that '1,100 who were of call-up age should be issued with notices to report for medical examination'. This threat served its purpose, and the apprentices, who were highly organised and motivated throughout, eventually returned to work on 12 April 1941.[51] The ramifications of the Interim Report, however, particularly the recommendation contained in item nine that apprentices should attend the subsequent negotiations, continued to worry the trade union leadership. In this regard, Gilbert, the secretary of the CSEU, reiterated his concerns to a Government official after the publication of a 'Second and Final Report' to Bevin had been made on 21 April. As Gilbert noted, 'we have no illusions as to the mind of the Court on this matter. The members of the Court must have known, from what had been quite frankly stated beforehand, that no Executive body could agree to such a course.' Gilbert went on to explain that the failure of the CSEU to accept the recommendation put the relationship between apprentices and union officials in jeopardy, and in his opinion it should not have been promised.[52] The conduct of the dispute in general had earlier prompted the Deputy Industrial Commissioner Scotland, J.B. Galbraith, to lament that he would welcome 'anything that would have the effect of undermining the almost fanatical devotion of the Employers' Federation to procedure'.[53] This was a sentiment that Parker would have no doubt agreed to, as he also noted that the dispute could have been restricted, if not averted, had more urgency been shown by the employers and trade unions involved.[54]

Subsequently, no other strike reached this scale in the industry, but stoppages of work, albeit of short duration, were commonplace. Employers provoked a one-day strike at a Clyde shipyard by locking out, for half a day, workers who had arrived late for work.[55] It was not only apprentices who were dissatisfied, as plumbers' boys and girls at Denny of Dumbarton went on strike for higher wages. The senior Scottish Law Officer, the Lord Advocate, took a dim view of this and subsequently a local sheriff imposed a fine or a ten-day sentence of imprisonment on the strikers, many of whom were between the ages of fourteen and sixteen. Although the effects of the Denny strike were later ameliorated, older workers continued to strike for higher wages, with 120 riveters at the Caledon shipyard at Dundee out for five days in March 1942, and 100 riveters at Lithgows at Port Glasgow out on strike for three weeks in July.[56] Other disputes continued elsewhere, with one strike on Tyneside in 1942 almost doubling the days lost to strikes in that year.[57] In Scotland, however, as Parker noted, there was twice the amount of prosecutions for strikes in comparison with England and Wales, but fewer individuals were involved.[58] The rash of strikes common to the shipbuilding industry from 1941 onwards was at least confirmation that those involved were continuing, to all intents and purposes, their own private war. This escalation in stoppages of work had occurred just two years after the industry had experienced the least days lost to strikes in its history, and continued to the end of the war. For what it is worth, bearing in mind the severe limitations of the following figures (*inter alia* they do not give either the numbers involved in individual strikes or their duration as a proportion of the total), the industry's record on strikes during the war years is suggested in Table 8.

Table 8: Stoppages of work in the shipbuilding industry, 1939–1945

Year	No. of strikes beginning in year	No. of workpeople involved	Working lost days
1939	39	4,300	37,000
1940	65	10,100	37,000
1941	147	27,300	110,000
1942	111	42,000	192,000
1943	196	32,000	137,000
1944	199	44,000	370,000
1945	186	27,700	143,000

Source: PRO, CAB 102/877.

Inman stated that in the interwar period and during the war, the highest incidence of strikes in the munitions industries as a proportion of the workforce employed occurred in shipbuilding.[59] The rights and wrongs of industrial disputes, however, are not always transparent, and although compulsory arbitration was required, this was rarely, if ever, quick to happen. Local shop stewards often took action without national sanction, and in the Boilermakers' Society, in particular, local officials exercised a higher degree of autonomy than was the case in other unions.

Apart from strikes, local yard committees also had powers to discipline persistent absenteeism, and from the middle of 1942 these powers had been varied.[60] From then on, an increasing number of workpeople were sentenced for absenteeism and bad timekeeping to fines or imprisonment. On the Clyde, to January 1942, there had been 42 convictions; however, by October 1943 the conviction rate had reached 50 in that month alone.[61] Inman noted that information on absenteeism, which some employers suggested had been fuelled by high earnings for overtime and Sunday working, was incomplete, but that the Clyde district employers did have a card index system of record. Although this was better than no record at all, it was nevertheless problematic, due to subjectivity in recording absenteeism.[62] Absenteeism and general slacking were serious problems, but due to the lack of a coherent national system of reporting, the extent of the problem is impossible to quantify. However, George Hall MP, the Financial Secretary to the Admiralty did attempt to sum up the position after a number of visits around the country at the end of 1942. Hall suggested that 85 per cent of men worked really well, 10 per cent could do better and 5 per cent had to be regarded as 'slackers'.[63] In a sense, Hall was essentially trying to put a salve on a long-running sore as slacking in the shipyards had been newsworthy for some time as a result of the early emphasis on repair and conversions, when naval personnel had increasingly come into contact with civilian workers.

Criticisms of absenteeism, strikes and slacking, such as those made by Admiral Sir Roger Keyes VC, caused a great deal of controversy as they were widely reported in the national press. One allegation repeated second-hand by Keyes was that 121 electricians were told to hide in a warship at a Clyde shipyard and keep out of sight for two weeks as there was no work for them.[64] This allegation was instantly investigated by the SEF, and had arisen as a result of a fire in a ship some eighteen months previously. As a risk of explosion existed, the men were told to stand off in view of the vessel being submerged.[65] In this climate, with men and women risking their lives at sea, allegations and counter-allegations were made on a regular basis, and one example from the labour side will suffice to indicate the level of the debate in the press. At a north-west shipyard, two trade unionists signed a statement criticising a naval captain who had held a party aboard his ship, which they alleged had held up important war work. The workmen claimed to have been taken off their duties to lower the legs and incline the back of a settee in the captain's quarters. At a time of rationing, the captain and his 80 invited guests consumed two turkeys, twelve chickens, eight salmon, two boxes of sausages and an iced cake containing 36 eggs![66]

In the wider compass of industrial relations in the shipbuilding industry, some trade unionists concerned by the inefficient use of skilled labour did offer their services to boost production at critical stages of the war. For example, in September 1941, at the Walker Naval yard on the Tyne, outfitting tradesmen concerned at production bottlenecks suggested that their skills should be used in 'flying squads' to visit other shipyards and therefore improve output on the Tyne. Delays in hull construction had kept these electricians idle for up to two months, even though they were paid to remain at the yard, an indication of the ongoing high-profit potential of naval contracts, and of their skill value to the employers. The men suggested that the proposed flying squads should be under the control of a Ministry of Labour official rather than the employers. This was a situation that the latter were loath to accept as it cut across their fundamental priority to preserve their skilled labour at all costs during profitable periods of construction. Moreover, there was no guarantee that if the employers let their men form these squads and allowed them to go they would get them back. It was also feared that the formation of squads would cut across the employers' freedom to hire and fire people of their choice, and would limit their control in their own yards. Accordingly, Tyneside employers rejected the suggestion, with their spokesman, Robin Rowell, of Hawthorn Leslie, speaking against the idea at the Admiralty, and this led to its rejection.[67]

Other shipbuilding employers maintained their skilled labour by more covert means. For example, Cammell Laird, at Birkenhead, employed twice the number of men employed by Vickers, at Barrow-in-Furness in Cumbria, whilst being engaged upon a similar amount of work. This situation prompted a Ministry of Labour official to decry the Birkenhead firm's obduracy in employing female labour to free up skilled manpower elsewhere when 2,000 females were registered for employment at the Mersey Exchange.[68] Across the country in the north-east coast district, employers and men alike were hostile to any importation of Irish labour, and as Inman noted, the industry there 'refused to have them at any price'.[69] Moreover, as late as 1942, there were no females employed in a productive capacity in the shipyards of the River Wear.[70] These reprehensible attitudes were not only confined to the shipbuilding industry alone, but are indicative of a disturbing malaise at its very heart.

Despite the myriad problems that had beset the industry in the early years of the war, mercantile output had improved from a below target total of 801,000 grt in 1940 to a total of 1,156,000 grt in 1941. The latter total was substantially achieved by the further injection of new labour, and for the rest of the war output remained above 1 mgrt. This was achieved despite the top priority given for labour and materiel to the aircraft industry in 1940 and 1941—a state of affairs which prompted the shipbuilders to complain that 'the fortunate possessors of these priorities were able to dispense with the necessity for planning their production in any sense, which involved the husbanding of resources'.[71] These attitudes were symptomatic of mounting frustration as shipbuilding and repair were not deemed to be essential industries for the control of labour until March 1941, when the Essential Work (Shipbuilding and Repairing) Order 1941 came into force.[72]

This Order coincided with hopes that after the initial concentration on small naval escorts, a more balanced programme of construction would ensue. The Pearl Harbor débâcle did, however, alter strategic priorities as the need for capital ships such as aircraft carriers was brought sharply into focus. Nevertheless, carriers, battleships, cruisers, and to a lesser extent destroyers took a long time to complete and tied up the resources, particularly in fitting out, of the larger naval yards for much of the war. As such, a large proportion of new construction and conversion continued to be based on vessels that were easier to build and convert. Priority had been given to minesweepers, trawlers, and vessels of the corvette type. The Admiralty, whose relationships with the larger naval yards were on the whole good, had nevertheless long suspected collusion over prices in the Warship Group yards. The situation, however, was somewhat clouded by the need to maintain output at high levels, and in so doing not to antagonise the shipbuilders unduly. As long as price collusion had remained an in-house Admiralty problem, it was confinable. Parliamentary scrutiny, however, had begun to bite from 1938 onwards and by 1941 the Admiralty had finally lost patience and accordingly cracked down on Warship Group profiteering. In doing so, the Admiralty really had little choice, given the severe criticism it had received from politicians over the lack of proper financial controls within its departments.

As Ashworth has observed (relying on the evidence presented in reports of the Comptroller and Auditor General, the Public Accounts Committee [PAC] and the Select Committee on National Expenditure of the House of Commons), in 1940 a scheme had been proposed to determine the actual costs of the private shipbuilders.[73] This was done in order to gain accurate guidelines in which to set future prices for warships, but was deferred until January 1941, in part because of the raising of Excess Profit Tax to 100 per cent, which had raised doubts about its necessity.[74] By February 1941, however, the shipbuilding industry had expressed strong disapproval of the Admiralty scheme. This led to a concession by the Admiralty, whereby if three shipbuilders received an order for the same class of vessel, then the costs of all three, rather than one, would be taken into account. Thereafter, 22 firms agreed to an Admiralty investigation, which began shortly afterwards and covered 32 warships, including an aircraft carrier, a battleship, cruisers and destroyers. Of these vessels, 27, or 84 per cent, had been ordered between 1936 and the outbreak of war, with the remaining five being placed shortly afterwards. In all, the 32 vessels were valued at £90,000,000. As a result of this investigation, the PAC subsequently noted the extent of excessive profit rates, especially those on submarines, which were over 70 per cent. Overall, the median profit rate stood at 27–28 per cent, and this prompted one MP to observe that the Director of Naval Construction (DNC) at the Admiralty had already been wiped out as an expert in costing, to which the DNC, Sir Stanley Goodall, replied: 'only in the matter of submarines'.[75]

Faced with this evidence, and to a large extent having been hoisted by their own petard, the warship builders agreed that they had in fact made excess profits and offered to waive claims for extras on all warships ordered from January 1939. After an unsuccessful attempt by the shipbuilders to convince the Admiralty to accept the former's own figures, Ashworth notes that the firms eventually waived outstanding claims totalling £2,250,000 on 169 vessels. These included 3 aircraft carriers, 3 battleships, 14 cruisers, 40 destroyers and 41 submarines.[76] Thereafter, the old system was abandoned, and two years later group settlements based on actual knowledge of costs at a recent date, and estimates of completion costs provided by the shipbuilders through the Shipbuilding Conference, had become the norm. Subsequently, the Conference, which had secretly included a levy of 1 per cent on the price of all contracts ordered since 1939, paid £365,000 to the Admiralty after its existence had been discovered.[77]

Although warship building, conversion and repair had overridden the need for a more balanced programme of construction, corvette supply, as Postan observed, 'did not at first respond to the urgency of demand', and output in 1942 and 1943 lagged behind what was expected.[78] This situation was exacerbated by the American inability to complete 100 out of 150 vessels on order due to a severe shortage of steel plate. The American need for escort vessels, however, was ameliorated by the transfer of ten British-built corvettes in 1942, and by the sale of 25 Canadian-built ships earmarked for the Royal Navy. As a result, by the autumn of 1943, only 28 British—and twelve American—built corvettes were added to the complement of the Royal Navy, a total that fell far short of expectations.[79] By the end of 1942, more manpower was employed on mercantile building than at any time since the start of the war. Despite this, however, labour remained relatively immobile between districts, although the labour supply had been helped by the absorption of unemployed shipbuilders and by the transfer of former employees from other industries. This had held back any widespread adoption of dilution, as had differences of opinion as to its nature and extent that had arisen between the Admiralty and the Ministry of Labour. Mercantile output had nonetheless risen to a rate of 1,400,000 grt per annum in the last four months of 1941, but naval building and repairs still retained the highest priority. Continuous Allied shipping losses had occurred in 1941 and 1942, and in the latter year some 1,664 British, Allied and neutral vessels, amounting to 7,790,676 grt, had been lost, of which 1,160 ships of 6,266,215 grt had been sunk by enemy submarines.[80]

With this level of losses, it followed that if any subsequent invasion of Europe was to be successful, then the Battle of the Atlantic, and the battle against the U-boats elsewhere, had to be won. Given the demands on the already overstretched capacity and available labour supply in the naval yards, this dual aim would be substantially aided by an increase in escort vessel construction in mercantile yards, and by a concurrent increase in prefabrication of hull sections and landing craft assembly in firms outwith the industry. By this stage, if these aims were to be realised, then all firms, particularly the smaller concerns which had on the whole suffered disproportionately in relation to the mixed naval and mercantile builders from the lack of investment in the inter-war period, had to be helped if production was to be improved. In this regard, only a substantial injection of money would compensate for the years of neglect, but only after Government investigations into the industry had taken place. Moreover, the Minister of Labour, Ernest Bevin, had already publicly expressed his mounting frustration with the obduracy displayed by shipbuilding employers on the question of the increased employment of women.[81] In the private SEF yards, the numbers of women employed from September 1939 to June 1942 had risen by approximately 500.[82] Bevin later made it plain to the shipbuilding employers that any future increase in the workforce must be met entirely from the recruitment of women.[83]

Before considering the impact of Government investigations in detail, it should be noted that the Shipbuilding Conference did at least try a measure of self-help by reopening two shipyards, which had been placed on care and maintenance by NSS. The first, the Low Walker yard of Armstrong-Whitworth on the Tyne, employed 505 workpeople in March 1941, nine months after it had opened to build partly prefabricated vessels.[84] The second, the former Swan Hunter yard at Southwick on the Wear, was only opened up after Sir James Lithgow had made an implied threat to the hitherto recalcitrant Wear builders that unless they agreed to reopen the disused yard then some of the skilled workers would be placed elsewhere.[85] This initiative was hardly a resounding success, and, although a higher proportion of women were employed at these yards, Buxton has noted that skilled labour was diverted from elsewhere, and that only seventeen vessels were completed at high unit costs.[86]

The problems of the shipbuilding industry, which were deep-seated and presumably not susceptible to quick-fix solutions, had been apparent from the beginning of the war. Given the history of the industry in the inter-war period, and the experience of Government officials to date, any contemporaneous investigation into it was likely to highlight its shortcomings. This proved to be the case when two investigative reports, the first chaired by Robert Barlow for the Minister of Production, which was subsequently reinforced by a Report to the Machine Tool Controller by Cecil Bentham, were respectively completed in July and September 1942.[87] Both reports were critical and suggested a raft of changes including further dilution, improved plant and equipment, and an extension of welding and prefabrication. Barlow found that there was insufficient collaboration between the industry at large and the Admiralty, and pointed to delays in approving plans. Moreover, he considered that there appeared to be more scope for 'simplification of detail and standardisation of fittings'. Barlow also noted that mercantile builders were critical of the Admiralty's lack of co-ordination with industry, and even between its own departments, which tended to act in isolation from each other. Another ringing complaint was that complicated design and additional apparatus resulted in substantial delays to the fitting-out period. In a three-page missive to the industry, Barlow came to the rather damning conclusion that 'a degree of complacency amongst all concerned permeates the whole field of production'.[88] The Warship Group of private shipbuilders felt duty-bound to disagree with this criticism, leading them to discuss future tactics before meeting with the Controller of the Navy to discuss the import of Barlow's letter.

Murray Stephen, of the Linthouse firm of A. Stephen, and Robin Rowell, of Hawthorn Leslie on the Tyne, recommended that a small committee should be formed from within the Group, but that it should not be chosen by the Admiralty, and that the 'Ministry of Production should be kept out of it'.[89] At the main meeting, the Controller informed the shipbuilders that he did not wish 'circumstances to arise which

Plate 3: New technology: the first all-welded destroyer under construction at J. Samuel White's of Cowes during the Second World War.

caused . . . other Ministries to throw bricks at the Admiralty, and that he would welcome anything that came direct from the shipbuilders to improve production'. In reply to the mercantile builders' criticism of the multiplicity of departments within the Admiralty, the Group informed the Controller that this was more of a problem for that sector, and that it did not bother them. The Controller then commented that Admiralty alterations and additions to warships was a sore point, whereupon Sir Charles Craven of Vickers Armstrong made his feelings known in no uncertain terms over the word 'complacency' in Barlow's report. Thereafter, the meeting was concluded in the Warship Group's favour. Again they had been allowed to put their own house in order and had succeeded in keeping other ministries from interfering in the industry.[90] It did not take the Group long to press home an advantage, as just five days later they had formed four zonal committees chaired by senior figures in the industry in response to Admiralty concerns.[91] Ostensibly, a zonal structure would improve co-ordination and provide a platform for discussion at the regional rather than at the national level, and would therefore be more attuned to local concerns. The Warship Group firms' real concern, however, remained the question of the Admiralty gaining detailed knowledge of their costings on particular contracts. The Group unanimously agreed that 'under no circumstances could shipbuilding firms' standard profit be upset'.[92]

On the whole the mixed naval and mercantile yards were far better equipped than were the mercantile yards. From the early days of the war, in part to compensate for the lack of suitable equipment in the latter, a measure of fabrication had been undertaken. A design suitable for tramp ship production based on the Lithgow Y type tramp steamer had been adopted by several yards. In turn, a new design known as the Partly Fabricated A type tramp ship had evolved. This was suitable for prefabrication on the basis of complete units weighing up to 7.5 tons. It is an indication of the overall lack of suitable cranage in many tramp yards that this design was never produced, and the structure had to be redesigned on the basis of smaller units with the capacity of the less well equipped yards. This led to the adoption of the Partly Fabricated B type tramp steamer which went into production at the beginning of 1941, of which 41 were subsequently built, before being superseded by another two versions. The more important contribution of partly and wholly fabricated mercantile vessels to the war effort is summarised in Table 9.

The problems of the cash-starved and spatially limited shipbuilding yards had to be tackled, and to an extent they were in the wake of the

Table 9: Partly and wholly fabricated hulls of mercantile vessels completed during the war

Type of Vessel	Nos built	% of steel weight in hull prefabricated
Tramps PF (B)	41	50
Tramps PF (C)	17	61
Tramps PF (D)	11	61
Tankers (Chants)	44	100
Coasters (Fabrics)	25	100
Coasters (Shelts)	24	100
Lighters	33	100
Tugs (Tids)	180	100
(A class)	15	100
(A class, Sterns only)	8	100
(B class, Burma)	36	100
Barges (Dumb, Burma)	54	100
Wagon Ferries	3	100

Source: PRO, CAB 102/440.

Barlow and Bentham reports. Although it has been previously stated that the industry, with some justification, was not susceptible to quick-fix solutions, this is exactly what it got, and amounted to, as Correlli Barnett noted, 'a remarkable feat of re-equipment in the middle of a world war'.[93] Bentham recommended that the shipyards and marine engineering establishments should receive 'exceptional financial consideration to enable them to deal with improvements in their plant'.[94] During his nationwide tour of shipyards and marine engineering works, Bentham noted that the percentage of modern plant on the north-east coast was around 25 per cent, and on the Clyde around 50 per cent. No doubt with an eye to more modern practice in America, Bentham recommended a rapid expansion of welding and associated plant, in conjunction with improvements in cranage and prefabrication.[95] Although it was undesirable to interfere with existing production, welding schemes were nonetheless encouraged, but many yards still relied on pneumatic or hydraulic riveting. With finance available, the Shipyard Development Committee (SDC) was established in November 1942 under the chairmanship of Sir James Lithgow, with Cecil Bentham in an advisory capacity. The SDC, to August 1943, oversaw an extensive programme of re-equipment of mercantile and naval yards. Lithgow sold the package of aid to the shipbuilders with the substantial inducement that the Admiralty would pay 50 per cent of the costs of the suggested schemes, particularly those involving welding.[96]

The extension of welding in British shipbuilding was the major and ultimately the most far-reaching change to the industry in wartime. Throughout the inter-war period, particularly with the restrictions on capital ships, the Admiralty had encouraged a gradual development of the welding process in the naval yards. In wartime, not least in light of the experience of the emergency yards in the USA, where all welded ships were turned out using techniques of multiple production, the Admiralty was keen to extend the practice.[97] From the outset, the SDC identified three areas of immediate concern in which modernisation was deemed to be necessary: the extension of welding schemes, the provision of new machine tools, and schemes for yard development including new and larger cranage. By September 1943, at which stage the tide of the Battle of the Atlantic had substantially turned in the Allies' favour, 90 per cent of the welding schemes instituted by the SDC were completed or were nearing completion.[98] Table 10 shows the extent of Admiralty expenditure on plant and machine tools for naval shipbuilding and marine engineering contractors in the five years from 1940 to 1944. Although it is likely that a proportion of these costs were subsumed into the costs displayed in Table 11, the overall picture on yard development and on welding schemes is nevertheless enhanced.

Table 10: Admiralty expenditure 1940–1944, plant and machine tools (£)

	1940	1941	1942	1943	1944
Expenditure	259,000	869,000	1,245,000	4,002,000	4,090,000

Source: PRO, CAB 102/442.

Table 11: Total costs of, and Admiralty contribution to, SDC yard developments and welding schemes in naval and merchant shipyards (£)

	Total cost	Admiralty contribution
Naval schemes:		
General development	3,084,618	2,490,482
Welding	1,399,669	916,397
Merchant shipyards:		
General development	1,671,599	1,162,956
Welding	776,866	451,781

Source: PRO, CAB 102/442.

As Table 11 shows, the total value of the SDC schemes was almost £7 million, of which the Admiralty provided just over £5 million. In welding, the Admiralty did in fact contribute more than the promised 50 per cent, but its contributions to general development in naval and merchant yards were greater, at 80 and 70 per cent respectively. Moreover, the need for expenditure by Government on this scale on yards that were generally unsuitable for any large-scale prefabrication schemes due to spatial limitations, was confirmation, if further proof were in fact needed, that the overall position of the industry was a sorry one.

By September 1943, the improved position in the Battle of the Atlantic, combined with the Axis defeat in North Africa, and the surrender of Italy, in tandem with the turn of the tide in the Soviet Union, redefined strategic priorities in preparation for the invasion of Europe. The prior emphasis on the construction of tank landing craft, the vast majority of which had been built by inland structural engineering firms, momentarily shifted back to the shipyards with a War Cabinet decision in November to build an additional 75 tank landing craft there.[99] Although this amount of craft did affect mercantile output, shipbuilding's overall contribution in this sphere was low, as by the end of the war 1,100 of over 1,200 tank landing craft had been constructed in structural engineering firms. Moreover, another 5,000 types of landing craft, including a substantial number mainly built in wood in boatbuilding yards, were completed.[100] The construction of tank landing craft in inland factories, some of which were up to 200 feet in length, by using techniques of pre-assembly and prefabrication, took a great deal of the strain off the shipbuilding industry, as did the inland prefabrication of escort vessels. Indeed, in the latter case, at least 80 per cent of the structure of the latest class of frigates had been prefabricated.[101] In this sphere, women played an increasing role, not only in inland factories but also in the boatyards where substantial numbers were employed.[102] One structural engineering firm, Sir William Arrol and Company, built all-welded landing craft at Kelliebank, in Alloa, Scotland. There, women had comprised around half of the workforce since 1942, and by May 1943 they had been increasingly used as welders.[103]

The rate of female dilution in the SEF yards had also increased by this stage, but the national picture was hardly uniform.[104] Thereafter, following years of inaction on the part of Government officials, shipbuilding employers and trade unionists, women played an increasingly important role in the shipyards until the end of the war. Nevertheless, labour shortages in shipbuilding persisted and by the autumn of 1943 manpower requirements in general were reaching a level that threatened an even higher degree of sacrifice than the considerable one already made by the civilian population. The Ministry of Labour Manpower Survey of October 1943 brought the situation to a head, as Ernest Bevin's covering memo indicated:

> The services are requiring . . . 776,000 men and women. The supply departments are asking for a net increase in munitions of 174,000, while the basic industries of the country (coal, agriculture, transport et cetera) are demanding an increase of 240,000. These demands cannot be met. Standards and amenities of the civilian population cannot be further reduced.[105]

The Prime Minister, Winston Churchill, adroitly summed up the manpower situation by stating that 'it was no longer a problem of closing a gap between supply and requirements, manpower could not be more fully mobilised for the war effort than it actually was'.[106]

With due respect to the efforts of other countries such as Australia, and the achievements of Canadian shipbuilding, which were considerable in what had previously been regarded as a minor shipbuilding nation, 1943 saw the high watermark of the huge American emergency shipbuilding effort.[107] By November, Colonel Knox of the United States Maritime Commission (USMC) had announced that more tonnage had been built than had been sunk in the entire war to date.[108] The largest emergency shipyard, Bethlehem Fairfield, with sixteen ways, built a total of 384 Liberty ships at an average construction time of 53.8 days each. The remarkable gains in construction times achieved in eighteen specially laid out yards that all used techniques based on mass production such as standardisation, sub-assembly and prefabrication easily surpassed British output times, but the American yards used far more labour. At the Oregon Shipbuilding Corporation yard, a total of 150 vessels were built in 30 days or less, with one built in fourteen days. This feat was equalled, and surpassed in one instance, by the No. 2 yard of Permanente Metals Corporation in Richmond, California, which built 157 Liberty ships, with one built in eight days.[109] By this stage, Britain, in desperate need of dollars to maintain its Lend Lease, arrangements had already sold to the USMC the two Todd Corporation shipyards it had owned since December 1940. This sale, at a rate of $4 to £1, resulted in a net loss to the British Exchequer of around £750,000. This was a transaction that the Treasury representative in Washington deemed, on the whole, to be favourable. In effect, Britain had transferred a liability, albeit for the Allied pool of ships, to build vessels in these yards to the USMC, a factor that prompted one parliamentarian to observe that Britain had sold her ally a 'pup'. Overall, the cost of building ships in the USA was around three times more expensive than in Britain, with the two yards costing around £7.5 million of £26 million spent on shipyards and on ships, out of a total of £39 million expended by the Admiralty on armaments.[110] Given the urgent need for these vessels at the time and subsequently, this was a relatively small price to pay, even though it had resulted in a considerable drain of much needed foreign currency. The spatially limited British mercantile and mixed naval shipyards could not possibly hope to compete with these production times. Moreover, from early 1942, production in some yards had been diverted into the building of merchant aircraft carriers (MACs), tramp ships equipped with a flight deck, elevators and a hangar. The keel of the first MAC was laid in August 1942, and some tankers also had a flight deck added to them.[111]

Nevertheless, although productivity comparisons cannot be relied upon, due to different cranage, plant, manpower and type of product, it is interesting to note the contents of Table 12, which provides a comparison of productivity figures of half a dozen or so naval firms in the USA, with those of their counterparts in the UK. Although it is not known just how representative these firms were, the comparison, in light of the disparity between comparable mercantile output, is illuminating.

Although building times were longer, in three of the four categories

Table 12: Productivity comparisons in output per man in tons per year in sample naval firms in the USA and the UK

	Building time (weeks)		Men on job tons per year		Output per man	
	USA	UK	USA	UK	USA	UK
Destroyers	42	92	635	207	4.1	5.5
Submarines	58	65	515	130	2.7	5.5
Prefab frigate	44	46	338	165	3.8	8.8
Carriers	–	–	2,590	765	8.9	6.2

Source: PRO, CAB 102/529.

of construction, the UK workforce, which was substantially less in numbers than its American counterpart, was more productive. Only in carrier construction, where the gap in numbers was greatest, did the USA surpass UK output. Although it is not known when this comparison was undertaken, the number of welders in the 27 main naval building firms in the UK had increased by 38 per cent in the twelve months to July 1943.[112] The expansion in welding was largely due to Admiralty contributions for improved plant, cranage and other equipment, and taken together with the expansion in prefabrication elsewhere is probably reflected in improved output. On the latter point, Sir Amos Ayre later claimed that average output had improved by 50–70 per cent per number employed at 1941–43 as compared to the period 1917–18.[113] This, in all probability, is an exaggerated claim; however, that it was made at all is indicative of the contribution made by the Admiralty in the re-equipment of the shipyards.

Table 13, extracted from Admiralty Appropriation Accounts, shows the extent of Admiralty expenditure on naval shipbuilding and other production from 1939 to 1944. These figures include armaments required for merchant vessels, though the cost of armanent production undertaken by the Ministry of Supply is not included after March 1940, nor is shipbuilding in the Royal Dockyards. As Hornby observed, Admiralty expenditure on armaments was never less than a third of total expenditure on naval requirements.[114] Shore establishments, the Royal Dockyards at home and abroad, and other Admiralty dockyards con-

Table 13: Admiralty expenditure for naval shipbuilding and other production, 1939–1944 (£m)

	1939	1940	1941	1942	1943	1944
Propelling and auxiliary machinery for naval ships	18.6	34.0	35.7	38.7	47.6	51.1
Hull construction	19.2	36.0	38.1	52.0	63.5	68.9
Total	37.8	70.0	73.8	90.7	111.1	120.0

Source: PRO, ADM 116/5555. British Admiralty Trade Mission to the USA.

sumed a large part of Admiralty expenditure, particularly in improving dock and berthing facilities. Even with the emphasis on repair and conversion, however, the Royal Dockyards were only able to handle just over one-third of all naval repairs, conversions and refits. Despite this, the three largest Royal Dockyards—Chatham, Devonport and Portsmouth—did manage to build 3 cruisers and 14 submarines during the war, against a total for the 1914–1918 war of 6 battleships, 14 cruisers and 29 submarines.[115]

It has been seen that mercantile output was at all times subject to changing priorities, and as Ayre noted the labour force was at no time equal to that of the plant and berth capacity of the industry, particularly on Clydeside.[116] As the probability of the war coming to a successful conclusion increased, it was likely that attitudes to dilution and the continued adoption of new technology from the established workforce would harden, as would those of management to the workforce in general. Regrettably, this generally proved to be the case, with one Government official, Mackie, noting in August 1944 that the Clyde employers had began to reassert their pre-war right to dismiss labour at will. Accordingly, a great many of the older boilermakers taken on during the war were now being 'thrown on the street'. Moreover, he also noted that the employers there were making a concerted drive to rid themselves of the obligations imposed upon them by wartime regulations, in order to return to the 'old starvation method of applying discipline'. Mackie then highlighted the perennial problem of labour relations in shipbuilding and repair by concluding that few if any employers realised that 'just as the power to discipline depended on the surplus of labour market, so soon did they lose that power when there were more jobs than workmen'.[117] It is likely, however, that there was an element of revenge on the part of the Clyde employers against their main antagonists in the Praetorian Guard of the shipyard trade unions, the Boilermakers' Society. At the start of the war, as Mortimer noted, the Society had a membership of some 60,000 men, a sizeable percentage of which worked in engineering and allied trades. When the membership increased by 50 per cent to reach its highest point of 90,000, this was 'proportionately less than some of the other unions in the engineering and shipbuilding industries'. Moreover, diluted labourers in the boilermaking trades were given temporary union cards for the duration of the war only.[118]

Given the boilermaking trade's position at the apex of the production process in the shipbuilding and repairing industries, it was hardly surprising that their attitudes to manning levels and further dilution hardened as the threat of national danger all but receded. In February 1944 boilermakers at Smith's Dock on the Tees were still firmly opposed to women brought into the industry as welders.[119] The position, yard-by-yard, however, was uneven; for example, at Swan Hunter in Wallsend, on Tyneside, boilermakers were still trying in March 1944 to obtain equal pay for women welders over the age of 21 who had completed their probationary period.[120] On the same river, at the Walker Naval yard, some 550 boilermakers, contending that only skilled men should work a new gas flame-cutting machine, went on strike over a management plan to entrust the working of the machine to semi-skilled labour. As the strike rumbled on into its seventh week in December 1944, the Boilermakers' General Secretary, Mark Hodgson, who later received a knighthood, blamed the employers for inciting the dispute by throwing down the gauntlet and appearing to be surprised that it was accepted. Moreover, if this attitude on the part of the employers was a portent of things to come, then 'all the talk of closer co-operation, harmonious relationship and mutual understanding becomes sheer hypocrisy, and prospects for the future are not bright'. Subsequently, despite strikes being illegal, and with a Court of Inquiry into the dispute already called, and an Interim Report into the dispute that favoured the employers, the men remained on strike. This led to some 125 men being fined £10 each by magistrates, with the alternative of 30 days imprisonment for non-payment. By June 1945, it had been reported that police in some cases were taking action to enforce the fines, which only nine boilermakers had paid. This provoked the Society's members at the yard to state that if the alternative of imprisonment for non-payment of fines were invoked, then the remaining workforce would refuse to work until they were released.[121]

As Hodgson had alluded to earlier, this type of industrial unrest did not augur well for the future of the industry. By this stage, in the wake of the Allied invasion of Europe, the employers had already met in the forum of their national organisations, the SEF and the Shipbuilding Conference, to discuss the post-war competitive position of the industry. At the initial meeting, Tristram Edwards, the President of the Shipbuilding Conference, pointed out that although the pre-war situation of the industry had not been good, the post-war climate would be substantially better. Nevertheless, he commented that the likelihood of a glut of orders should not hide the need to restore the industry's competitive strength in the light of its steady decline in the inter-war period. Edwards recognised that costs were at the heart of the competitive position, and that the goal of higher productivity, which would result in a reduction in the selling price of the product, had to be reached. However, this laudable aim would only be achieved through better co-operation between higher management and the trade unions.[122] Sir Amos Ayre's jaundiced view was that war output had been limited in direct proportion to the effort expended by the boilermakers, and blamed them for the industry's decline. By this stage, however, even Ayre realised that the industry 'must at least talk to the unions', and that they must be told (as if they did not know already) what 'exactly was the position of the industry in 1938'.[123] Thereafter, following much discussion, John Boyd of the SEF stated that 'in the long-run the test was what the industry was prepared to do, not what the unions were prepared to do'.[124] In this regard, the shipbuilders were well aware that if the Labour Party came to power after the war, it could not rule out the possibility of nationalisation of the industry. Moreover, it would stress the need for full employment and its wish to embrace the unions as partners in the drive for exports to replenish Britain's vastly depleted foreign currency reserves.[125]

After further consideration by the SEF and the Conference, a second meeting took place in October, wherein a Committee on Improved Shipbuilding Practice was constituted, and a Sub-committee on Methods of Shipbuilding Construction was formed as an adjunct.[126] Subsequently, the latter spawned four other sub-committees that looked into more detailed aspects of shipbuilding, but nevertheless decided to issue an Interim Report before the other sub-committees had fully reported. The report, issued in March 1945, bore the imprimatur of Sir

Amos Ayre, and repeated many of his earlier assertions.[127] It did, however, appraise the threat to the industry's competitive position when international competition resurfaced, and left it up to all concerned 'to reverse the pre-war competitive relationship of Continental and UK yards'.[128] However, the attainment of this ideal rested on the co-operation and the collective will of the employers to see it through. In this regard, before the Interim Report had been discussed, the Clyde shipbuilder Maurice Denny saw the urgency of the need for the industry to change its ways, stating that: 'Unless our industry takes this seriously and goes into the future with a complete determination on a 'must' basis, we shall bequeath to our successors the same legacy of strife, frustration and comparative stagnation that has been on the whole a characteristic of our industry in the past.'[129] Denny's prescient concern that the shipbuilding industry collectively had to accept and foster the need for change, had to be pitted against the extant system of work organisation, which remained casualised and offered no guarantee of permanency of employment, nor interchangeability between crafts. Moreover, the post-war position, when the USA would effectively withdraw its vast emergency shipbuilding resources from the market, and the disarray of its major continental competitors, offered the British shipbuilding industry an opportunity not only to retain but also to advance its pre-war position as the world's largest shipbuilding nation, in what effectively was a protected market.

In this regard, Government re-equipment of the yards in wartime had been a distinct advantage, but perversely this investment probably held back further dilution in the industry, rather than encouraging it. Similarly, there is a dearth of evidence to suggest that the employers used diluted male and female labour to de-skill occupations previously undertaken by skilled men. In the last years of the war, however, many yards, particularly in the boilermaking trades, suffered from shortages of labour, especially Clydeside where some yards were only half-manned on a single shift throughout the war.[130] Dilution of the workforce in yards where welding had been adopted eased the path of new entrants into the steel working trades. Riveting, however, remained the principal means of metal-joining in many mercantile yards. Its future, despite the advance of welding in the war years, seemed to be assured, as the Conference and Works Board of the SEF noted in May 1945 that 'there is a definite future for riveting and recruitment should in general be by means of apprenticeship'.[131] This example of employer conservatism in not fully embracing new technology—although there is natural tendency to persist with the extant methods of work organisation, plant and equipment that had served firms well in the past—hardly faced up to the new construction methods developed in the war. Moreover, there was substantial evidence to suggest that apprentice labourers were beginning to shun riveting as a trade. As Inman noted, throughout the war there was no difficulty in getting apprentice labour for welding, electrical work, shipwrights and joinery; however, there was 'a great problem in getting boys to take up riveting'. Furthermore, as riveting was a boilermaking trade, the Boilermakers' Society had insisted from the nineteenth century onwards, without any agreement from the employers, that the ratio of tradesmen to apprentices in riveting should not exceed 1:5.[132]

What was, however, in many senses an unprecedented level of co-operation between management and labour during the war meant that the Admiralty and the Ministry of Labour on the whole left the private yards to continue much as they had done before. Despite this, however, the private shipbuilding and repairing industries' contributions to the war effort were considerable, although the country did also rely on shipbuilding output from Allied countries elsewhere. Tables 14, 15 and 16 show mercantile and naval tonnage output and types of naval vessels built in the private yards during the war years. Owing to differences in the measurement of tonnage, these figures are not comparable. Figures for mercantile tonnage are given from 1940 to 1944, as these totals are more reliable. Taking into consideration the longer fitting-out period of warships, the years 1939 to 1945 are considered, as are the types of naval vessels built.

Table 14: Merchant tonnage, new construction, vessels completed, 1940–1944 (000 gross tons)

	1940	1941	1942	1943	1944
Tonnage	801	1,156	1,301	1,204	1,014

Note: In terms of tonnage, apart from 1940, over 1,000,000 gross tons was completed despite the priority given to naval construction, conversion and repairs in every year to 1944. And, overall, from January 1940 to August 1945, 6,000,000 gross tons of merchant shipping was completed.

Source: NMM, Shipbuilding Conference Statistics.

Table 15: Output of naval vessels in the UK, 1939–1945, excluding vessels built outside the shipyards (dt)

	No. of vessels	Tonnage
Large naval yards	674	1,365,430
Small naval yards	590	468,517
Large merchant yards	44	60,032
Smaller merchant yards	36	26,300

Source: W. Hornby, *Factories and Plant* (London, 1958), p. 46.

Table 16: Types of naval vessels built in the shipyards, 1939–1945

	Large naval yards	Large merchant yards	Small naval yards	Small merchant yards
Battleships	5	–	–	–
Aircraft carriers	14	–	–	–
Cruisers and minesweepers	28	1	–	–
Destroyers	225	–	–	–
Submarines	143	–	–	–
Sloops and minelayers	36	2	–	–
Corvettes and frigates	55	13	194	8
Fleet minesweepers	33	8	41	8
Depot ships and misc.	6	1	66	2
Landing ships	20	4	3	1
Landing craft	39	15	16	–
Coastal forces craft	70	–	–	–
Trawlers	–	–	204	15
Boom defence	–	–	55	–
Motor minesweepers	–	–	13	–

Note: Excludes bulk of coastal forces craft which were built in boatbuilding yards, and the bulk of landing craft, which were built outside the shipyards, and the three cruisers and fourteen submarines, built in the Royal Dockyards.

Source: W. Hornby, *Factories and Plant* (London, 1958), p. 47.

In terms of tonnage, apart from 1940, over 1 mgrt was completed despite the priority given to naval construction, conversion and repairs in every year to 1944. And overall, from January 1940 to August 1945, 6 mgrt of merchant shipping was completed.[133] Clearly the bulk of the supply of tonnage came from the larger established naval yards, some of which constructed the capital ship component of the figures. Table 16 gives the type of naval vessels built in the shipyards.

The total of 1,344 naval vessels built between 1939 and 1945 inclusive, which in terms of complexity, length of outfitting and skilled labour used far outweighed the mercantile output of 1,576 merchant ships launched during the same period, was on the whole a creditable performance. This is all the more so if we take into account the lean years of the inter-war period, when much of the industry's capacity lay unused and neglected. Overall, the output of the British maritime industries in general, as Buxton has commented, 'together with its later allies was just sufficient to maintain the Royal and Merchant Navies, but not by a large margin. The supply of labour and materials was always the limiting factor.'[134] German U-boats also sank 175 Allied warships, the majority of which were British, and were, as Roskill has noted, 'a source of anxiety right to the end'. However, with over 81,000 lost lives in the Royal Navy and Merchant Marine as a result of enemy action, Roskill got to the heart of the matter by succinctly asking the question: 'must the same inadequacies again be redeemed at the same price in lives?'[135] The war had at least rejuvenated the British maritime industries, and offered them the opportunity to steal a march on their competitors in the post-war period. Indeed it would not be going too far to argue that the Second World War had saved the industry in the opposite sense that the First World War had almost ruined it. It remained to be seen, however, just what the competitive response would amount to, before Britain's main continental competitors, including Germany, eventually re-equipped and re-entered the world market for ships. Taking advantage of the entirely new conditions presented by the post-1945 period would require considerable shifts in mental attitude. Could British shipbuilding take advantage of the opportunities presented?

4

The Missed Opportunity

The Second World War certainly had the effect of restoring the UK shipbuilding industry to profitability. Many of the UK's leading competitors in the inter-war years had been knocked out of realistic competition. The British shipbuilding industry, therefore, was presented with an opportunity to re-establish itself in world markets in terms which had not existed since before the First World War. The 1950s would, however, become something of a 'devil's decade' in terms of the shipbuilding industry. It would bask in the easy conditions of the late 1940s and early 1950s, but when the cosy world of cost-plus building and a seller's market gave way to serious competition with fixed prices, and a buyer's market in the later 1950s, British shipbuilding began to fall apart. New trade routes, new trades, new construction methods, the decline of liner traffic and the rise of the airliner all hammered traditional British attitudes and approaches to the market. Still, the decade offered huge opportunities to British shipbuilders; that they could not take them said much about the conditioning of the past. That shipbuilders abroad were not so hidebound spoke volumes.

What was to be unique about the 30-year period following 1945 was that it was marked by almost continuous economic expansion and concomitantly a huge increase in the volume of seaborne trade. World and individual fleets expanded as never before and consequently so did the demand for ships. The world merchant fleet expanded from 29,340 vessels totalling just over 80 mgrt in 1948 to 41,865 vessels totalling over 160 mgrt in 1965. The UK's merchant fleet in 1948 comprised 6,025 vessels totalling just over 18 mgrt but by 1965 the number of vessels had fallen to 4,437 vessels, although tonnage had expanded to 21.5 mgrt. As the world fleet doubled, therefore, Britain's only grew by 16 per cent, while its percentage of the world's fleet fell from 24 to 13 per cent. Much of this was eloquent testimony to the bespoke nature of demand in British shipbuilding.

Consideration of the post-war position of the shipbuilding industry began in 1942.[1] Little was achieved, however, and by 1944 the two ministers responsible for control of the industry, A.V. Alexander, First Lord of the Admiralty, and Lord Leathers, the Minister of War Transport, both believed that some form of Government control of the industry in the post-war period was essential if shipbuilding were to avoid what Alexander termed the 'chaotic conditions' of the past.[2] Both men were agreed that the experience of the industry in the inter-war period had bred highly conservative attitudes which were a barrier to future efficiency.[3] As had been the case during the First World War, however, the industry wanted the Government to withdraw as quickly as possible. At a meeting to discuss the regulation of orders, the Joint Committee of Shipbuilders and Shipowners (JCSS) made its views clear through J.R. Hobhouse, the Chairman of the General Council of British Shipping:

> The position is this . . . [that we] are . . . fairly convinced that we shall do the job better than the Government will do it . . . I will not conceal from you that we would very much like to be free of Government control and I think . . . we should take the view that we ought to fight to get rid of Government control.[4]

In response, Alexander and Leathers, pointed out that the recent history of the industry proved that it was incapable of solving its own problems and that Government assistance was essential. As Alexander put it, there was 'no use leaving the shipbuilding industry to drag on as it was left after the last war'.[5] If anything, however, the attitudes of the shipbuilders were proving even more intractable in 1944 than they had been in 1917.

Just how difficult such attitudes could be was given an early airing by the President of the Shipbuilding Conference, A. Murray Stephen, in 1944. Responding to Government proposals on full employment, Stephen used the occasion of his presidential address to the Institution of Engineers and Shipbuilders in Scotland to discourse on the subject 'Full Employment in British Shipyards'. Reviewing the industry's recent past he stated that:

> No industry has had such a record of booms and slumps in the past as British shipbuilding . . . How can any industry function satisfactorily when it is alternatively subjected to famines and feasts . . . ? History has shown quite clearly that wars have had . . . a very

> adverse effect on British shipbuilding, and in particular the war of 1914–18 brought in its train a set of circumstances which brought the industry in this country to such a pass that its chances of survival on any large scale were of the slenderest.

The result of the post-1918 boom, according to Stephen, was 'an industry expanded to 133 per cent of pre-1914 capacity, having available . . . less than half its pre-1914 demand'. As to the future, and in order to prevent a repetition of the inter-war years, Stephen advanced the following programme: the Axis powers should be prevented from shipbuilding and their merchant fleets either prohibited or subjected to controls; certain trade routes and all coastal routes should be reserved for British shipping; all British registered ships should be constructed in British yards; flagging-out should be banned and there should be no post-war disarmament with naval orders being applied contra-cyclically. Only through the full implementation of such measures, Stephen concluded, could the ravages of the inter-war period be avoided and full employment in the shipbuilding industry be guaranteed.[6] As a set of policy descriptions, this amounted to a veritable Treaty of Versailles from British shipbuilders, would have gladdened the heart of the most committed of mercantilists and, in all probability, struck dread into the hearts of Government ministers.

Still, at least in the interim, the Government got something of its own way. Alexander and Leathers advanced the view that Government control was necessary, at least in the transition period, but that it did not need to be rigid, nor be a precursor to any full-scale inquiry into the industry. What they envisaged was an independent, well-informed body which could anticipate any difficulties which the industry might face and duly alert the Government to them. What was proposed was a shipbuilding advisory board to act as a conduit between Government and industry; with both sides accepting the recommendations, the Shipbuilding Committee was established in November 1944, with its operational life extending to early 1946.[7] It oversaw a system of permits to shipowners, allowing the regulation of priorities between owners, ship types and yards. The licensing system remained in force whereby the Admiralty would issue a licence to enable materials to be ordered when it had received a permit. The aim of the permit-licence system was to enable the Government to keep control of the industry should demand outstrip capacity. By July 1945, the issuing of permits to neutral owners was approved and by the year end it was decided to discontinue the permit system. The final recommendations of the Committee were that the British mercantile marine should be restored to its pre-war figure of 18 mgrt and that the level of employment in the shipbuilding and ship-repairing industries should not fall below 90,000. Its final conclusion was that by 1955 world tonnage would stand at 70 mgrt, UK tonnage would be 18 mgrt and that in the same period the average annual output of merchant ships in the UK would be 1.25 mgrt per annum.[8]

When the Shipbuilding Committee lapsed, Alexander took the view that a permanent body to oversee the development of the industry should be established. In the opinion of the Lord President's Committee:

> there is every advantage in announcing this body and getting it to work quickly so as to leave no cause for uncertainty in the industry and at the same time to implement the warning already given in your announcement on nationalisation that the industry, although not down for nationalisation, will be expected to have full regard to the public interest . . . an independent Chairman . . . in the circumstances . . . seems inevitable, but the appointment may be a precedent for other industries which are not, or not at present to be nationalised.[9]

Probably cowed by even the prospect of nationalisation, the shipbuilders duly acquiesced in the formation of the Shipbuilding Advisory Committee (SAC) in June 1946. The SAC, however, quickly became a 'talking shop' captured by its representatives, the shipbuilders, the shipowners and trades unions, rather than an organisation which could, in Alexander's words, 'plan the longer term future of the industry'. It rapidly degenerated, therefore, into a forum for the airing of vested interests.[10] The early meetings of the SAC were dominated by shortages of materials, particularly steel and timber, and also of skilled labour, particularly in the finishing trades. The trade press also reported material and labour shortages but, in a worrying echo of the post First World War situation, noted rising prices as a major problem. The prices of British-built ships were believed to be between 70 and 100 per cent above the those of 1938. The price of steel plates had risen from £10 10s 6d to £16 16s 6d; labour costs had risen some 75 per cent and the price of completed accessories was up by nearly 200 per cent.[11] Despite noting that there was 'little likelihood that prices will fall' and that 'the tendency may well be in the opposite direction', the mood amongst shipbuilders remained buoyant: 'because of the urgency of the need for new tonnage, the present high level of building costs has perhaps played a secondary part in recent orders, early delivery dates most of all being keenly sought'.[12] The first quarter figures for 1947 revealed that the industry had the largest volume of work on hand since 1922, and it was believed that such an improvement would be maintained throughout the year. Worrying signs were, however, noted with the Chairman of Hawthorn Leslie, Robin Rowell, taking the view that 'it was disturbing . . . that virtually no deep sea ships of importance were being equipped to burn coal'. There was also anxiety over the future of those yards which had formerly depended upon the production of tramp tonnage as their staple output.[13]

Once again, the attitudes of British shipbuilders to the market proved worrying. There was a general feeling that the market trend was away from specialised, coal-fired tramp production and towards standardised, diesel-propelled tanker production as the staple of the industry, with all the implications which this entailed for production methods and yard layouts. The shipbuilders were quite correct, the trend was away from coal-burning tramp vessels towards motor-powered tankers and other specialised tonnage, but nothing was done about it. Such a stance allowed British shipbuilders both to worry about and to decry foreign competition at the same time. American methods of producing standardised ships in prefabricated sections, for example, had, 'never . . . seriously disturbed or dismayed' British shipbuilders because 'their confidence born of assurance in the technical skills and craftsmanship, was never more justified than it is today by the present healthy state of the industry'.[14] As N.M. Hunter, the Chairman of Swan Hunter and Wigham Richardson, commented in 1947:

> I think there is a great danger of the spectacular building of a very simple type of comparatively small cargo ships of one design, built in the USA during the last war years in great number, creating an impression that the methods used were new and so far ahead of any in this country that they should be slavishly copied.[15]

All of this displayed a dangerous arrogance, for though American competition, given the cost differential, could reasonably be ignored, the fact that methods and layouts could be copied by almost anyone does not seem to have been considered. Indeed, whilst German and other European capacity was regarded as being 'out of the picture for many

Plate 4: Swan Hunter *c.*1946. The clutter of the warshipbuilding programme is obvious but note the covered berth and the floating crane.

years to come', when the SAC received a progress report on the dismantling of German capacity, it reacted strongly, with Murray Stephen declaring that the situation 'confirmed the industry's worst fears in regard to demolition . . . as . . . practically nothing had been done to dismantle the German yards'. Echoing this theme, Sir Amos Ayre maintained that 'if the United Kingdom were to maintain its pre-eminence Germany could not be allowed to enter into competition'. The Committee further took the view that representation should be made to the First Lord that 'it was of the first importance for the British shipbuilding industry . . . that the steps for the demolition of the German yards should be carried out'.[16] All would be well, therefore, as long as no foreign competition were permitted.

This fear—there can be no other word for it—of the likely post-war market was compounded in that 'most of main British shipbuilders obtained half or more of their orders . . . from a relatively small and clearly identified band of shipping companies who ordered with them'.[17] Over many years most British shipbuilders had sought to stabilise their market positions by forming alliances with buyers, and a bespoke builder–client relationship had developed, largely based upon the personal relationships between the chairmen and managing directors of the building firms and shipping companies. The striking feature of this situation was not its strength but rather its weakness, in that the shipbuilders had little power to influence the market, which was driven by the owners. The best that could be said was that shipbuilders knew with some accuracy what the preferences of their customers were. Thus, both the domestic and overseas order book tended to reflect the concentration of activity in a few well-established market areas. British shipbuilding, therefore, was a well-schooled industry; it constructed to order and had little regard for professional marketing in terms of ship types and likely market trends. Despite its predominance in the immediate post-war period, therefore, the industry remained highly segmented, indeed almost atomised, in its structure, in which bespoke conditions and defence of individual market shares had become normal. Heavily conditioned by the inter-war era, few British shipbuilders saw the need to change their minds in the post-1945 period.[18]

As is demonstrated in Table 17, Britain, once again, failed to meet the challenge of foreign competition in a rapidly expanding market. Although one should be wary of making too much of these statistics as one commentator has observed, a declining national share of a rising world market for an internationally traded product, such as ships, even

Table 17: World and UK shipbuilding launches, 1947–1962, and UK as a percentage of the world (mgrt), 1947–1962

Year	World	UK	UK as % of the world
1947	2.1	1.2	57
1948	2.3	1.2	51
1949	3.1	1.3	41
1950	3.5	1.3	38
1951	3.6	1.3	37
1952	4.4	1.3	30
1953	5.1	1.3	26
1954	5.3	1.4	27
1955	5.3	1.4	28
1956	6.7	1.3	21
1957	8.5	1.4	17
1958	9.3	1.4	15
1959	8.7	1.3	16
1960	8.4	1.3	16
1961	7.9	1.3	15
1962	8.3	1.1	13

Source: *Lloyd's Register of Shipping.*

where the absolute level of national production is not falling, can usually be taken as an indicator of deteriorating national competitiveness. In some senses, a falling national volume with an increasing international market share is preferable to an increasing national volume and shrinking international market share, since the latter usually portends serious problems at the next downturn in the trade cycle. The experience of the British shipbuilding industry fits broadly within this description. Whereas absolute levels of output can be sustained for some time if world demand rises faster than new capacity, if either world demand falls or new capacity outstrips it, an uncompetitive industry, or shipyard, will be the first to feel the impact, and probably the last to recover, given that orders will go to the more competitive sectors first. The only way in which this situation can be avoided is through increased competitiveness.[19]

The best that can be said of the data in Table 17 is that British output remained remarkedly steady over the period, but this was in the context of a vibrant expansionary market and reveals a stark trend of relative decline. The weakness of the bespoke nature of British shipbuilding is further revealed in Tables 18 and 19. The fall in overseas registrations relative to domestic registrations is clear and the degree of import penetration equally obvious. With domestic demand dwindling the failure in the export market is even more startling, as evidenced by Table 19. In less than fifteen years, in a market of unprecedented strength, Britain had lost both international market share and had its domestic market penetrated to a considerable extent. Alone of the major maritime nations, Britain's relative share of the world fleet declined substantially as the world fleet doubled. It was a failure, in a dynamic and expansionary market, of dramatic proportions. It can be explained by the failure of the British shipbuilding industry to adapt itself to what might be termed a second industrial revolution.

In many respects this situation was underpinned by trends in market preference and ship type which had been established in the inter-war

Table 18: UK shipbuilding, 1948–1962 (000 grt)

Year	Total	Domestic registration	Overseas registration	Imports
1948	1.2	766	410	0
1949	1.3	745	522	0
1950	1.3	884	441	0
1951	1.3	739	602	0
1952	1.3	888	415	21
1953	1.3	953	365	49
1954	1.4	927	482	54
1955	1.4	935	539	37
1956	1.3	949	435	119
1957	1.4	1,152	261	158
1958	1.4	1,064	388	285
1959	1.3	1,257	115	449
1960	1.3	1,185	146	465
1961	1.2	911	281	509
1962	1.1	908	165	569

Source: *Lloyd's Register of Shipping.*

Table 19: Shipbuilding: percentage shares of the world export market, 1948–1950, 1951–1955 and 1956–1960

	UK	Japan	West Germany	Sweden	France	Netherlands	Others
1948–50	35	2	0	18	0	6	38
1951–55	22	11	15	13	2	9	29
1956–60	7	32	21	12	6	6	17

Source: *Lloyd's Register of Shipping.*

period. In terms of ship type, two trends had dominated the inter-war years, the growth of the motor ship and the rise of the tanker, and in both of these developing markets British shipowners and builders had proved sluggish. At the end of the First World War, a tiny percentage of the world merchant fleet was propelled by diesel engines, but by 1939 nearly one-quarter was thus propelled (nearly 17 mgrt of shipping). Indeed, motorships comprised some 36 per cent of all tonnage launched in the inter-war era. British shipbuilders built 28 per cent of their tonnage as motorships in this period, while for foreign builders the proportion was 42 per cent. British builders moved more slowly than their foreign counterparts towards the diesel engine, in large part reflecting a substantial historical investment in the marine steam engine and the centrality of coal as a fuel. It was also a testament to the importance of coal as a commodity in the coastal, short and long sea trades.[20] Shipowners in the inter-war period also appeared reluctant to specialise, in that once a diesel engine was installed, the ship was reliant upon oil. Doubts were also expressed about the oil/coal price ratio. As the RMSP Chairman, Lord Kyslant, put it: 'there are two facts . . . which are tending to hold back the more general adoption of motorships in place of steamships—namely the uncertainty of obtaining, and at reasonable cost, the necessary supplies of oil; and secondly, the present relatively high first cost of motor engines compared with steam engines'.[21] Given the growth of the motorship, however, Robin Rowell's view on the virtue of steam was particularly worrying.[22]

A similar position pertained with respect to the development of the tanker between the wars. Britain contributed to this growth, and throughout the 1920s tankers comprised about one-sixth of both British and foreign tonnage. During the 1930s, however, tanker tonnage accounted for one-third of the foreign shipbuilding market, while in Britain it represented less than one-quarter of launchings, and the total launched actually declined. As British registered tanker tonnage increased from 1.7 to 2.7 mgrt, non-British tonnage more than doubled to 8 mgrt. Again, this in part reflected the central role of coal in the British economy and its importance as a cargo, but it also revealed a deep-seated, if totally ironic, fear of over-specialisation. The slow move towards tanker construction reflected the bespoke nature of British builder–client relationships, heavily wedded to the liner, cargo liner, coastal and short sea trades. Throughout the post-1945 period, the huge demand for tanker tonnage was treated by British builders as something of an abnormality. For example, in 1953 the trade journal *Fairplay* quoted a 'leading shipowner' to the effect that 'the peak for this class of tonnage (tankers) has been reached', while the shipbuilding correspondent of the *Shipbuilder and Marine Engine Builder* reported that 'the tank ship market has lost its golden halo'.[23]

In many respects, the sluggishness of tanker building was matched by the slow adoption of welding technology, which would later bestow upon other countries, notably Japan, Sweden and West Germany, a competitive advantage in crude and bulk carriers and tankers. The first all-welded vessel was built in Britain by Cammell Laird in 1920, although the concept was slow in diffusion. Both the Admiralty and *Lloyd's Register* doubted the stress-resisting capacity of the lighter welded hull. Despite the fact that tests in the USA in 1929 had demonstrated a 21 per cent saving in terms of costs, an 18 per cent saving in terms of labour and a 16 per cent saving in terms of weight of welded over riveted construction, surprisingly little interest was shown until the development of the submerged-arc process in the mid-1930s.[24] It was the spur given to tonnage in the Second World War, particularly through the Liberty ship and T2 tanker programmes in the USA, which brought the technology into its own. In the first round of Liberty ship construction, vessels were built on average in 149 days, whereas in the last round the average was down to 41 days. What was gained in productivity, however, tended to be obscured by high costs and poor quality control, there being a number of spectacular ship failures. The basic problem was not welding *per se* but the fact that design details commonplace in riveted structures led to fatigue initiation in their welded counterparts. This was compounded by variable quality steel, which tended to behave in distinct ways in certain conditions, in the worst case being prone to brittle fracture.[25] Although this was eventually overcome, it probably did enough to confirm the virtues of riveting in the minds of many British shipbuilders. As the representative of the Director of Naval Construction's department commented in 1942, 'it is the DNC's policy to increase welding as much as possible, and . . . the present position is that we cannot get as much welded structure as we would like . . . firms should be encouraged to weld rather than rivet'.[26] Little of this, however, discomforted British shipbuilders.

Conservative attitudes to the market on the part of the British shipbuilders in the late 1940s abounded. For example, the Chairman of the Wearside yard Joseph L. Thompson, Cyril Thompson, could comment on his yard's full order book, which included tankers, that 'in order to build these vessels, which are practically all-welded, we have been forced to put on order . . . capital equipment . . . costing £28,500'. The language here is instructive: 'forced' to spend £28,500 over two years, little more than double the directors' fees for the same period, against profits of over £160,000 for the year ended March 1948. But Thompson continued:

> You will appreciate that while we have a very good order book we must face the fact that this expenditure is necessary to hold our own in a highly competitive market which calls for the building of mainly welded ships. The shipbuilding industry is notorious for seven lean years following seven fat ones and bearing this in mind, we hope to be able to establish substantial reserves including a dividend equalisation reserve.[27]

Thompson's position is interesting on a variety of levels, identifying as it does, as early as 1948, clear market trends, the rise of tanker orders as a proportion of the total order book, the concomitant rise in all-welded construction and a steady increase in the size of ships. The response to the rising market however, was to prepare for the next slump!

Nor was this an isolated view, as the Chairman of the Teeside builders Smith's Dock, T. Eustace Smith, put it:

> There is no question in my mind that the shipbuilders in this country have been slow in modernising their yards. During the 1914–18 war a certain amount of alterations were made and this should have been continued after the war. As you will remember, we ran into boom periods and people were so busy making such good profits that they could not afford the time and dislocation of output to make

> improvements. We eventually ran into a slump period and then no one felt they could afford to do any modernising of their yards . . . I am certain that this is absolutely vital if the shipbuilding industry in this country is going to compete with the 'foreigner' when the slump period comes.[28]

This was the problem in a nutshell: yards could not invest in a boom because of the dislocation of production and nor could they invest in a recession because of the lack of capital. Such views could stand for the chairman of almost any British shipbuilding company: respond piece-meal to the current market conditions, never forget that the next slump is bound to come, stiffen the reserves to get through it, and protect the dividend at all costs.

Even so, the industry received a nasty shock in 1947 when the Blythswood yard at Scotstoun on the Clyde lost a Norwegian tanker order to Sweden because of rising costs and inability to guarantee the delivery date.[29] One year later a 'Glasgow shipowner' was advancing the view that:

> Shipowners, for the first time since the war, are sensing signs of a falling off in the urgent demand for ships. It may be only a lull, it may be temporary, or it may be the start of a slump. In past years, after boom periods, once there was an excess of tonnage a slump in freight rates followed rapidly . . . During boom periods shipowners, being human, are easily led astray and become over-optimistic and extravagant and build ships which ultimately prove too expensive for their trades. That happened in 1918 and is likely to happen again.[30]

This doom-laden prediction was all the worse because, by the same owner's calculations, world capacity was still 8–10 mgrt below its 1939 level. In March 1948 the Ellerman Line cancelled orders for six vessels because of costs and delivery dates, and both the Shipowners' Association and the Chamber of Shipping petitioned the Government to do something with respect to delivery dates and costs.[31] The situation was bad enough to provoke the Government to mount two separate lines of enquiry. First, the Labour Party's Research Department conducted an analysis of the industry, and then the Shipbuilding Costs Committee was established to examine the issue of shipbuilding costs.

The Labour Party's Research Department produced its report in

1948. This outlined the main problems of the industry as being the bespoke nature of demand and the consequent fluctuations in building, the lack of any marketing strategy, geographical concentration and consequent problems of unemployment in periods of slump, and demarcation mainly caused by fluctuations in employment and poor labour relations. The report proposed that any plan for the industry should have as its aim, the securing of industrial efficiency and, as far as possible, the maintenance of full employment. Any such plan had to produce a total annual capacity proportionate to the needs of the British merchant fleet and the export market, achieve a smooth flow of launchings to utilise maximum capacity, maintain stable employment and 'achieve . . . the highest standard of technical efficiency'. On the final issue the authors of the report felt that financial assistance could well be necessary. The report agreed with the consensus view of shipbuilders that 'the slump comes as surely as night follows day', and advocated a detailed scheme that would eliminate redundant capacity in such a way that the remaining yards had sufficient work and avoid what had occurred between the wars. The shipbuilders themselves had failed to do this in the inter-war period:

> partly because they were largely occupied with the financial aspects of the case . . . paid . . . too little attention to questions of efficiency and unemployment . . . and . . . partly because firms would not be persuaded to abandon their individualistic attitude and forget their own particular interests in the interest of the industry as a whole. It would seem, today, that some form of nationalisation might best achieve that unification of the industry which is vital for the success of any attempt to draw up a plan for the industry as a whole . . . a scheme which embraces the entire industry is essential if any lasting prosperity is to be achieved.

The report argued that new methods of construction, utilising welding and prefabrication, required a complete reconstruction of the yards, and that this could be best achieved via Government finance and assistance. Although alternatives to nationalisation were considered, the report concluded that there was 'no . . . good reason . . . why the State should act as nurse maid to private enterprise when it is in trouble. It seems inevitable that a prosperous shipbuilding industry will require heavy expenditure and it is very doubtful whether the industry is willing and or able to undertake this.'[32] The Government decided, however, that with the industry's order books more or less guaranteed until 1951 it would simply keep the situation under review.

The original intention behind the Shipbuilding Costs Committee had been that the Minister of Transport should have a series of talks with shipbuilders and shipowners over costs, but he had been persuaded otherwise by the Permanent Secretary of the Ministry of War Transport, Sir Cyril Hurcomb, on the grounds that it was a private matter between the shipowners and builders. Having removed the Minister from the equation:

> the shipbuilders then did their utmost to influence the terms of reference of the Committee as to make it ineffectual in dealing with the two principal matters with which the Minister was most concerned, namely, why were shipbuilding costs so much higher than pre-war and what the level of costs were to make it possible for our shipbuilders to compete with the foreigner? In the end it was agreed that the Committee would have nothing to do with the total price of ships . . . They were precluded from taking into account shipbuilder's profits and the Ministry was very careful to stand aside from their activities and to avoid being drawn in as witnesses for the Committee.

The end result had been 'a very non-controversial report', which provided 'no useful recommendations on which action could be taken'. The Committee, dominated by shipbuilders, and taking evidence from shipbuilders, concluded that there was nothing wrong with the geographical layout of the yards, nor with the management, financial strength, rate of modernisation of the industry and, most blatantly of all, the cost structure of the industry. Summing up the report, the Ministry of Transport commented that:

> there is nothing wrong with the industry that increased supplies of materials, fixed quotations from the sub-contractors and increased output from labour will not cure. The pronouncements are stated as conclusions but the Committee make little or no attempt to illustrate the evidence from which they draw their conclusions . . . the Report offers little help in determining whether shipbuilding costs are right or wrong or whether the industry is in need of overhaul or not. It neither calls for nor suggests further action.

In the view of the Ministry the report was so useless that 'the best thing to do . . . was to . . . find some way of burying it'.[33] Once again, however, the Government view was that with the industry having 4 mgrt on its order books, any extension of control would simply deter shipowners from ordering and it was therefore 'of the utmost importance . . . not to come to the assistance of the industry too early and thereby interfere with the powerful incentive to reduce the prices of new ships'.[34]

Despite the huge amount of tonnage on order, the shipbuilders continued to carp about the nature of the order book. G.W. Barr, Managing Director of Fairfield and a past President of the Clyde Shipbuilders Association (CSA), bemoaned 'the complete lack of enquiry for the highest classes of passenger and cargo liners, for coasters and smaller craft generally'. The heavy flow of tanker orders merely served to conceal these 'disturbing features', which were 'clear enough to those who appreciate the importance of balance as well as volume in the shipbuilding order book'.[35] Tankers, therefore, were useful enough for ensuring the continuity of work, but somehow did not represent 'real' shipbuilding. The then current president of the CSA, Charles Connell, agreed, stating that:

> many yards have never built and are not designed or equipped to build tankers, although in the face of the great tanker demand some yards have undertaken this class of work for the first time. Obviously the tanker boom brings no comfort or orders to the many builders of small craft, coasters, colliers and specialised river and estuary vessels. Tankers provide, proportionately to other types, less work for the finishing trades, which are so important in the building of high-class passenger vessels.[36]

Tankers were simply not 'class' vessels. Even in 1951, with the slipways satiated by the Korean War boom, having an order book comprising 1,000 ships, of 6.25 mgrt, worth £550 million and representing four years work, the President of the Shipbuilding Conference, J. Ramsay Gebbie, could still complain that it was 'an abnormal and artificial order book'.[37] With tankers comprising half of all orders in 1951, George Borrie commented on the improving spread of work and Murray Stephen publicly congratulated the shipbuilders for having 'kept their place in world output'.[38] This hardly reflected the reality of the situation, however, in that the percentage share of UK launches in the world had fallen by 11 per cent in only four years and the shipbuilders appear to have settled for a steady state approach to the market, launching an annual average of 1.3 mgrt between 1947 and 1953.

By 1953, however, with the order book glutted, launchings going back, and costs rising accordingly, two related worries began to be aired by shipbuilders: the rise of foreign competition, particularly from Germany and Japan, and the prospect of British orders being placed abroad.[39] Such concerns eventually reached the ears of the elderly Prime Minister, Winston Churchill, who demanded a report. This sparked a joint reply from the Admiralty and the Ministry of Transport which was a curious mixture of the confident ('the building berths in the major shipyards are fully booked until 1956') and the concerned ('we could not afford to lose our position as the leading shipbuilders and shipowners of the world. If there were any such threat, the Government would have to take drastic steps . . . to assist these industries').[40] More worryingly for the Government were the comments of Lester Goldsmith, the Chairman of the Gulf Oil Refinery Company, which offer a revealing insight into the prevailing mind-set of British shipbuilders. Goldsmith had approached the industry with a view to having two tankers built in Britain and was 'disgusted' with the response from the merchant shipbuilding sector. He was told that he would have to accept low temperature and steam conditions; he would have to accept packing in the condenser tube plates and that rolled tube ends would not be considered; in fact he would just have to accept what the shipbuilders decided to give him. As the Admiralty observed:

> according to Goldsmith the 'brush off' was always ready with: 'our order books are filled'. He felt very strongly that, 'It was when your order books are filled that you can afford to do something different'. He placed orders for two tankers in Belgium, the turbines being built under licence to Westinghouse design.[41]

For an industry which had long prided itself on building exactly what the customer wanted, no matter how specialised, the attitude displayed not only conservatism but an insouciance bordering on the dangerous. Furthermore, given that Goldsmith's views were part of an Admiralty review of the 'present weakness of the marine engineering industry and its consequences', the consequences were potentially far-reaching.

Rising costs and late delivery dates now became the most frequently aired criticisms of the industry. Thomas Falck, the Chairman of a Bergen shipping company, observed that Norwegian shipowners had placed orders for 35 ships of approximately 260,000 grt abroad between 1953 and 1954, but none had been ordered in Britain because 'British shipyards were not competitive today'.[42] Given that Norway was traditionally Britain's most lucrative and largest export market, this was a particularly worrying trend. Britain held over 30 per cent of the Norwegian market between 1948 and 1955 but thereafter demand all but collapsed and there would be much agonising over the loss of the Norwegian market in British shipbuilding circles.[43] There was no doubt that the major problems afflicting the industry were rising costs, late delivery dates and lack of acceptable credit facilities. Sir William Souter, the Chairman of the Sheaf Steam Shipping Company, stated that an 18,600 dwt oil tanker, ordered in February 1950 at an estimated cost of £735,000, had been delivered in November 1953, seven months late, at a cost of £949,817.[44] Given that such vessels were ordered with a view to the spot and time-charter markets, the potential loss to the shipowner went well beyond the actual cost of the vessel itself. Once again, a worrying ignorance of the modern shipping market was beginning to reveal itself. As the tanker boom was enhanced following the closure of the Suez Canal in 1956, the trade press would comment, somewhat dolefully, that this 'emphasised the capacity of yards abroad to build to modern requirements and sizes . . . [and that] owners generally are less inclined to be impressed . . . by the superiority of British workmanship and design'. British shipbuilders, however, had an ingenious answer to such criticisms. As one trade journal opined, the 'conservative policy', with regard to modernisation, productive capacity and the market place, would prove to be 'wise', since with the passing of the boom any disadvantage which accrued from late delivery dates would rapidly disappear. This, it should be noted, was in the same article which reported a record tonnage, 377,120 grt, being built abroad for British owners, and the fact that foreign yards were building more for British owners than British yards were building for owners abroad.[45] Indeed, the mild recession which set in during 1958 only served to vindicate the view of the shipbuilders that 'the seven lean years' had at last arrived.

It was the lack of British success in the tanker market, particularly in the wake of the post-1956 Suez boom, that finally allowed full penetration of the British ship market and hammered its export markets. As Table 20 demonstrates, import penetration only became serious in this market after 1956. Three trends are obvious: first, like the rest of the industry, steady state production for the domestic market; second, a trend fall in production for overseas registration; and, finally, the inor-

Table 20: Tanker launches UK, 1950–1959 (000 grt), and tonnage imported as a percentage of total launched

Year	Total	Domestic registration	Overseas registration	Imported
1950	614	367	247	0.0
1951	834	340	494	0.0
1952	643	353	290	2.8
1953	762	527	235	4.9
1954	714	424	290	2.8
1955	648	354	294	0.0
1956	518	333	185	4.6
1957	535	474	61	22.2
1958	577	396	181	35.2
1959	529	479	50	72.9

Source: *Lloyd's Register of Shipping.*

dinate rise in imports from 1957. With shipowners no longer willing to accept cost-plus quotations and the market increasingly moving in favour of the buyer, the whole cost structure of the British shipbuilding industry was condemned for the second time in less than 40 years. In April 1957 Sir James McNeill, the President of the Shipbuilding Conference, wrote to the First Lord of the Admiralty, Viscount Hailsham, concerning the shipbuilding statistics for 1956 and the first quarter of 1957. As McNeill pointed out, 'for the first time in peace-time history, the United Kingdom has had to take second place to a foreign power, viz, Japan . . . [and that] all-time record launchings were established by Germany, the Netherlands, Italy and Norway'. In McNeill's view, the figures indicated 'a most definite comparative trend', and that shipbuilding was facing a crisis.[46] In the following year, the Warship Group Chairman, Murray Stephen, met with the Controller of the Navy and two points were clear: 'firstly, that the Programme was so meagre that I hoped he would do his best to see that all the crumbs possible came our way . . . . Secondly . . . that history seemed to be repeating itself and that a slump in the merchant side of the industry looks like coinciding with a slump in naval work.'[47] This view was reinforced when, in turn, McNeill argued that 'at the time when merchant activity seems likely to fall off very considerably, the naval programme is providing very little cushion of compensating activity'. Indeed, surveying the last 50 years of shipbuilding history, McNeill purported to prove 'how this unfortunate feature of the fall off of naval work has coincided with the bad merchant years'.[48] Yet again, a post-war naval construction programme had failed to provide any relief to the industry, but unlike the post-Great War situation, on this occasion neither disarmament nor depression were to blame.

Much of this stemmed from discussions between the Admiralty and the shipbuilders in the 1940s, when, under the aegis of the war emergency, it had become something of an article of faith between the two sides that neither naval power nor the shipbuilding industry should ever again face the traumas of the inter-war years when a combination of disarmament, naval limitation treaties and the collapse of international trade had resulted in the near emasculation of the British shipbuilding industry and the consequent problems which emerged in both the rearmament period and during the Second World War. As a Cabinet paper made clear in 1945, 'it was most important from the point of view of the Government's employment policy that full weight should be given to the possibility of varying warship building in order to even out the slumps and booms of merchant building'.[49] The acknowledged aim, therefore, was to use naval shipbuilding in a contra-cyclical fashion to even out the boom–slump cycle. It was recognised that this phenomenon had done much to exacerbate the difficulties of the inter-war years and had contributed to serious skill and capacity shortages, as well as heavily localised unemployment. What the Government wished to avoid was the lack of investment which had characterised the inter-war period and the enforced reduction of capacity managed by the industry itself. This strategy, although laudable in and of itself, had a variety of flaws. The post-Second World War period did not resemble that of the post-Great War period in that international trade, rather than collapsing, boomed. Thus, for the first full post-war decade orders for the mercantile side of the industry dominated the order books. This, in its turn, was compounded by the uncertainties surrounding the Admiralty's construction requirements in terms of both numbers and types of ships, as it sought to develop a distinctive role in an uncertain international environment, attempted to come to terms with rapid technological change, and had to operate within a budgetary environment typified by economic stringency. Moreover, the pattern of Admiralty requirements, if pattern is the correct term for eclectic ordering, was difficult, at least initially, to fit into the glutted programme of mercantile building.[50]

In an examination of the balance between mercantile and naval shipbuilding, it was recognised that it was necessary 'to have long-term ship construction programmes, with sufficient flexibility to allow for warship building to be most active in periods when merchant shipbuilding is low and vice versa, assuming that considerations of defence, etc., permit this course'.[51] This was exactly what neither post-war defence requirements nor mercantile demand allowed after 1945. Much of the thinking was conditioned by the post-Great War situation of brief boom followed by long-term slump. In fact, the strength of the international economic recovery from 1945 was probably beyond reasonable forecasting and the reconstruction of the mercantile marine was to last, sustained by the Korean War commodities boom and the tanker boom consequent on the first Suez Canal closure, until the late 1950s. Given the full, and lengthening, order books of British shipyards in the 1940s and 1950s, it proved very difficult to synchronise warship building with mercantile building. The difficulty for the Admiralty and the Royal Navy, however, resided in exactly what it was building to defend against. The major problem which the Royal Navy faced, in common with the other services, was that of settling on a clear strategy in a period of rapid technological change. The advent of the atomic bomb in 1945, followed by the hydrogen bomb in 1953, all coupled with the developing Cold War, transformed the traditional preoccupations of the Royal Navy. Was it to prepare for 'Hot War' or for waging the 'Cold War'? Was it to concentrate on preparations for nuclear war, including 'broken-backed warfare'; to concentrate on limited warfare on the Korean War model; or to concentrate on the Cold War through an extension of the previous imperial practice of 'showing the flag?'. As the Admiralty Board noted in January 1955:

> The succession of Ministerial reviews of the rearmament programme . . . has resulted in the plans for Naval New Construction in particular never having been stable for long enough to permit the preparation of detailed submissions. They have been progressively whittled down in step with our financial prospects . . . the 1953–4 and 1954–5 programmes are the remnants of previous larger hopes, particularly so for frigates, and as they now stand cannot be regarded in themselves as the product of any conscious planning and indeed appear to be open to criticism in this regard.[52]

It was unlikely, therefore, that any assistance could come to the industry from a hard-pressed and confused naval sector.

With the Government seeking defence cuts of between £150 and £200 million in 1955–56, the Admiralty came under consistent Treasury pressure to affect a rationalisation of the Warship Group. Following the enquiry into marine engineering, which had concluded that the effective monopoly of the marine engineering firms had to be broken if Admiralty requirements were to be met, the Deputy Engineer-in-Chief of the fleet concluded that there were only two 'progressive' shipbuilding companies in the UK, Yarrow and Cammell Laird. While a number of firms recognised the need for modernisation, the report's author commented that 'few have the vigorous management necessary to do anything about it'.[53] This, in turn, led to a joint meeting between the Treasury and the Admiralty which concluded that there was a need for a wholesale modernisation of the shipyards in terms of layouts, plant and equipment particularly to increase the use of prefabrication. The return on capital employed and investment rates were criticised as poor and British costs were felt to be comparatively high in international terms. Restrictive practices were noted, but more worryingly there was a shortage of skilled labour in the industry, which was responsible for the British single-shift system, with consequently expensive overtime, compared with the double-shift system which was normal on the Continent and in Japan.[54] Against this background, the First Lord of the Admiralty, Lord Hailsham, told the Minister of Defence, Anthony Head, there could be no future cruiser programme and even a future aircraft carrier programme was doubtful, as he dolefully predicted 'the end of our world-wide naval power'.[55]

With the Admiralty attempting to move from cost-plus tendering to fixed costs, the reaction from the industry was predictable. The Shipbuilding Conference duly warned the Admiralty that 'while the warship building programme remains so emasculated and so frequently is only able to proceed on slow motion lines the Admiralty must not expect to get anything like the same low rates of profit terms as have applied when they were the predominant users of the shipbuilders facilities'.[56] This was ill-timed, however, as with the onset of the mild recession the seller's market gave way quickly to a buyer's market. Despite the pressure on defence cuts the main aim of the Admiralty and the Cabinet in 1959 was still to use naval orders to alleviate unemployment black spots. The Admiralty, however, noted that there were 'few items which can be commenced now', and suggested that the current building programme of one cruiser and eight frigates, which had been slowed down the previous year through budgetary constraints, should be restored to the original completion dates. This would, the Admiralty believed, 'result in increased employment, in the yards concerned . . . [although] the amount of increased employment is difficult to determine without detailed investigation but the increase at peak would amount to about 1,000 men'.[57] This revealed the difficulty of attempting to use naval orders to increase employment with nine vessels generating an employment gain of 1,000 men distributed around seven shipyards. By 1959, with the buyer's market beginning to bite, the Economic Policy Committee of the Cabinet termed the position 'disturbing' and felt that it was 'questionable whether we . . . [could] retain even our current percentage of world shipbuilding . . . [and that] the industry would begin to encounter acute difficulties in 1960'. Despite the usual suggestion that the Government should examine the possibility of phasing orders to assist the industry, there was a developing view that the real solution was that the industry had to 'contract and re-organise'.[58]

The criticisms of the Admiralty and the Treasury over modernisation and investment certainly carried weight. Vickers-Armstrong was one of the few yards to invest in reconstruction and extensive modernisation in the immediate post-war years, spending £2.25 million. At the other end of the scale, Cammell Laird did begin an £18 million modernisation scheme, but only in 1956.[59] Investment rates in shipbuilding, however, need to be compared with those in the manufacturing sector in general.

As Table 21 shows, the investment record in shipbuilding compared unfavourably with manufacturing as a whole, and with some individual sectors, dramatically so. It is significant that the figures for shipbuilding do begin to rise towards the end of the period when, arguably, the best chance of cashing in on the expanding market had already been missed. As one economist noted at the time:

> The striking thing about the British ship[building] industry . . . is its stubborn refusal to invest in the creation of more productive capacity . . . the amount of money spent on plant and equipment can have been barely sufficient to cover normal wear and tear and obsolescence in the shipyards . . . British shipbuilding firms spent £4 million annually on their fixed assets. For an industry which was producing an average of £120 million a year . . . this is a figure which is so low that it would suggest to the outside observer that someone was trying to get out of the business, and in the meantime was determined to spend as little as possible on it.[60]

Table 21: Capital investment as a proportion of net output in shipbuilding and other industries in the UK, 1951–1957 (%)

Year	1951	1952	1953	1954	1955	1956	1957	*Average*
Sector:								
All manufacturing	9.2	9.4	8.7	8.7	9.7	11.1	11.6	9.9
Mechanical engineering	8.2	8.8	8.2	7.2	8.6	9.0	9.4	8.5
Electrical engineering	5.6	6.3	6.0	5.9	8.2	9.3	8.6	7.3
Motor vehicles & cycles	8.8	8.3	7.1	7.4	13.0	18.4	18.8	12.2
Metal manufacture	14.3	12.4	12.2	15.0	13.0	15.2	18.4	14.5
Shipbuilding & repairing	4.1	3.7	4.2	4.4	4.1	4.7	6.4	4.6
Marine engineering	6.6	7.2	7.4	7.9	8.6	5.8	7.2	7.2

Source: PRO, BT 291/54, Shipbuilding Advisory Committee, Sub-Committee on Prospects, 22 July 1960.

This was not far off the mark in that in a few years' time many yards would seek voluntary liquidation as a way of getting out of the industry.

There was, however, considerable force in the criticism, especially when investment rates are compared with profit rates and distributed dividends. Between 1945 and 1959, a representative sample of thirteen firms paid average annual dividends of over 10 per cent, distributing over £2.6 million per annum. In other words, thirteen firms were distributing in dividends over half per annum what the whole industry was spending on its fixed assets. This approach, though, seemed to satisfy the shipbuilding shareholders and the markets. As Table 22 demonstrates, whilst shipbuilding performed poorly in investment terms, it clearly outperformed all other manufacturing industries in terms of share prices.

Analysis of the accounts of one company, Yarrow's on the Clyde, between 1945 and 1955 can be taken as a case in point. The accounts show that the company had a good deal of cash, with total liquidity rising from £236,710 in 1945 to £1,312,820 in 1955. Net profits rose from just under £101,000 in 1947 to almost £557,000 in 1953, before falling back to just under £400,000 in 1955. The company paid out almost half its after-tax profits in dividends in 1947, declaring a dividend of just under 43 per cent, paid its lowest dividend of the ten-year period in 1952, 16 per cent on after-tax profits of £191,192, and declared a 24 per cent dividend on after-tax profits of £184,839 in 1955. Movement on share capital in the same period was entirely due to making capitalisation/

Table 22: Share price index for manufacturing

Sector	End 1949	October 1956
Aircraft	100	209
Brewing	100	113
Building	100	134
Chemicals	100	204
Electrical engineering	100	140
Engineering	100	185
Foods	100	163
Motors	100	160
Plastic	100	111
Radio and television	100	230
Rubber	100	157
Shipbuilding	100	288
Textiles	100	140
Tobacco	100	86

Source: *Financial Times*, 13 October 1956.

bonus issues, the purpose of which in the late twentieth century would be to lock up distributable reserves by making them permanent. While the motive is not immediately obvious in the Yarrow case, the intention seems to have been to maintain or increase the later dividend. In terms of investments, the figures reveal that land and buildings investments rose from £65,000 in 1945 to £230,000 in 1955, but plant and machinery increased only from £167,000 to £200,000 over the same period. Of far greater importance than internal investment were stakes in subsidiary companies, government stocks and temporary investments, such as tax reserve certificates, with government stocks and temporary investments accounting for over £1.3 million in 1954–55. While one could debate the typicality of Yarrow's, its financial structure—family controlled with a fairly narrow shareholding base—was not so far removed from other shipbuilding concerns to destroy its validity as an example. Some key points are worth reiterating: the company was making healthy profits, producing strong dividends, was cash rich, whether in liquid form or in terms of outside investments, but appeared remarkedly reticent with respect to internal investment.[61]

The industry, therefore, was not exactly strapped for cash, but the investment record during the years of sizeable expansion and high demand was very disappointing in both absolute and relative terms, and testifies to a very conservative approach to the market. Despite relentless complaints from the shipbuilders about their figures in the post-war years, the gripes were, essentially, hollow. As the trade press noted as early as 1947, 'profits have shown a rising trend over the past twelve months and there have been many and quite substantial increases of dividends . . . even so, there has been ample margin left as cover for distributions so that already strong finances have been further strengthened'.[62] At the other end of the chronological scale, the Committee of Inquiry into Shipping which reported in 1970 and also took cognisance of shipowners' concerns, dismissed the complaints in somewhat brusque fashion:

> While there is undoubtedly room for argument on the relative financial position of owners in different countries over the period of war and early post-war recovery . . . we can find no evidence that scarcity of finance, either from existing reserves or on reasonable terms from outside sources, exercised any necessarily limiting influence on the development of the UK fleet.[63]

The complaints, however, were made in the context of a huge increase in competition for the British mercantile marine. The competition had been growing before 1914 and had become intense in the inter-war period. The ultimate impact, of course, was to threaten to undermine the bespoke nature of the UK shipbuilding/shipping relationship. The world merchant fleet doubled from just over 80 mgrt in 1948 to 160 mgrt in 1965 and reflected the absolute growth of nearly all mercantile marines in the period. Table 23 illustrates the main trends, and points to a number of conclusions. First, between them, fourteen countries held over 86 per cent of the world merchant fleet in 1948, and that same fourteen, with the addition of Liberia (a fact which illustrated the growing importance of flagging out), held over 83 per cent of the world merchant fleet in 1965. The second obvious feature is the share of growth rates for almost all individual nations. In some cases, Liberia, Japan, Norway and Greece, this growth was spectacular and resulted in substantial increases in market share; in others the growth was less spectacular, if just as marked, but the only countries to lose significant market share were the USA and Britain.

Table 23: Merchant fleet growth, 1948-1965 (000 grt), and percentage shares of world fleets (in brackets)

	1948		1965	
World	80,292	(100)	160,392	(100)
Denmark	1,223	(1.4)	2,562	(1.6)
France	2,786	(3.5)	5,198	(3.2)
West Germany	300	(0.4)	5,279	(3.3)
Greece	1,286	(1.6)	7,137	(4.4)
Italy	2,100	(2.6)	5,701	(3.6)
Japan	1,024	(1.3)	11,971	(7.5)
Liberia	–	–	17,539	(10.9)
Netherlands	2,737	(3.4)	4,891	(3.0)
Norway	4,261	(5.3)	15,641	(9.8)
Panama	2,716	(3.4)	4,465	(2.8)
Spain	1,147	(1.4)	2,132	(1.3)
Sweden	1,973	(2.5)	4,290	(2.7)
USSR	2,097	(2.6)	8,258	(5.1)
UK	18,025	(22.4)	21,530	(13.4)
USA	29,165	(36.3)	21,527	(13.4)

Source: *Lloyd's Register of Shipping.*

With the arrival of the buyer's market, all aspects of the British shipbuilding market—domestic, overseas and naval—came under increased pressure. As already observed, there was little hope of any alleviation of such pressure from the naval sector, as the Ministry of Transport conceded in 1960, commenting that 'the pattern of employment on ships building on present plans for the Royal Navy will not change drastically during 1960–1961 and will not significantly alter the picture of the future based on merchant work only'.[64] Taking advantage of the advent of the buyer's market, the Admiralty moved to competitive tendering with fixed prices being, as the Treasury noted 'the spur to build with the greatest efficiency and economy'.[65] The whole idea of attempting to use naval orders in a contra-cyclical fashion was dropped, although mopping-up individual employment difficulties was not. Instead, a new and more hard-headed policy was now revealed:

> Some of the less efficient firms who have come to rely on Admiralty work to an important extent, can be expected to press for the placing of orders for Warships to continue on the old basis . . . . Among these are some old established firms with long records of building ships for the Admiralty. Resort to competitive tendering may therefore be criticised as inconsistent with the aim of tackling local pockets of unemployment: the invitation of tenders would not, however, bind the Admiralty to accept any particular offer, so that local unemployment problems could still be taken into account in allocating orders. On the other hand it is no longer Government policy to retain a reserve potential for an emergency, and the number of warship building firms has in fact become an embarrassment, in as much as the naval programme these days is not large enough to make it possible to give all of them a continuity of orders.[66]

Indeed, the Treasury went even further, insisting that any area in which a warship could be built was almost by definition an area of high unemployment and therefore it made little difference where contracts went.

The impact of the late 1956 Suez crisis provoked considerable debate on oil supplies to Britain which fed in to the wider discussion on British shipbuilding. The Government established a committee under Treasury chairmanship to consider the transport of oil from the Middle East which, given the nature of the transport, was effectively a committee on shipbuilding. Here, once again, the first issues to be raised were the late delivery dates and high costs of British shipbuilding, as a consequence of which 'British owners in increasing numbers are having to build in foreign yards'.[67] As the Admiralty and Ministry of Defence viewed the situation:

> the acute world-wide shortage of tankers is such that it may be necessary for British shipowners to build in United States yards where the costs may be upwards as twice as in other yards (*sic*). Price incentive is thus not lacking for British shipbuilding managements to expand . . . It is thus most desirable for the health of the industry that shipbuilding output should be increased, otherwise British shipbuilding, already overtaken by Japan, may well decline still further through loss of goodwill and consequent failure to recapture markets once lost . . . unless delivery dates are improved an increasing number of British as well as foreign orders are bound to

> be lost to other countries where shipbuilding industries are prepared to expand.[68]

The final report of the Working Party on the Transport of Oil from the Middle East was produced in mid-1957. The report noted that 'the British shipbuilding industry has been comparatively sluggish in responding to the opportunities from expansion offered by the growth in oil and other traffic'.[69] The report criticised the industry's investment record and highlighted the employers' memories of the 1930s as a major inhibition to the expansion of capacity, although it did comment favourably that seven yards had modernisation schemes afoot. The Committee reasoned that the world output of ships would average some 7 mgrt per annum—but again, and in common with most forecasts this was an underestimate as output averaged over 9 mgrt per annum in the period—and it suggested that British capacity should be expanded to 2 mgrt per annum to restore 'the British share in world output'.[70] The Committee believed that an output of 2 mgrt per annum was feasible by 1961, although this represented 'the maximum which the industry would be willing to contemplate', and that 'we think it would be prudent to avoid forcing the growth of the industry by powerful artificial stunts'. The somewhat limp conclusion was therefore that no intervention was required in a situation which was best left to 'commercial judgement', but that there could be 'discussions with the shipbuilding industry' to ensure that 'they are not under-rating the full possibilities of expansion'.[71]

Unfortunately for this report, its dissemination coincided with the downturn in the shipping market and shipbuilding. As a consequence, when asked in 1958 what progress had been made in terms of increasing capacity and output, the Admiralty's comment was that 'the shipbuilders do not find themselves in an expansionist mood', despite the fact that the Admiralty believed that there would still be a 'substantial world demand for new tonnage', and the suggestion was advanced that more attractive credit terms should be offered to encourage both domestic and foreign orders.[72] By the end of the year the Treasury had become so concerned about the industry to suggest that a comprehensive paper should be prepared for the Cabinet as 'the industry was on the verge of a most serious depression'.[73] The report by Treasury officials appeared in July 1959. The conclusions were bleak: the industry's output had remained static, its share of overseas orders had declined and increasingly British shipowners were ordering new ships overseas. The industry, the Treasury observed 'is not competitive. It has not modernised its production methods and organisation so quickly or so thoroughly as its competitors. Its labour relations are poor, and its management, while improving, is not as good as it might be.' The report did concede that some modernisation was proceeding but that 'in no case has the plan been as bold as that of the best yards of Sweden, Germany and Japan . . . [and share] a lack of enterprise and foresight'. The report was pessimistic over delivery dates and prices and concluded that the prospects of the industry becoming competitive could not be rated very high. Nonetheless, it advanced a range of options: a Government enquiry into the state and prospects of the industry or a full-scale independent enquiry; the establishment of a high authority for the industry with powers over capital investment, research, rationalisation and labour relations; and financial or fiscal inducements to shipowners to stimulate shipbuilding. All of these, however, either implied, or were predicated on, Government financial assistance and 'if Ministers were prepared to concede the possibility of such help, it would be necessary to make it clear . . . that Government assistance could only be considered on a strictly temporary and short term basis'.[74] As one official was quick to note, 'action on these lines would commit the Government to financial help . . . the difficulty is whether to let the industry rip, or take some expensive and very dubiously prospectful action to put it back on its feet. Ministers will have to decide this before they meet the industry.'[75] The Treasury, it seemed, viewed decline as inevitable.

The Treasury, however, could settle on neither a strategy to put to ministers, nor any recommendation as to when the industry should be approached. Its thinking seemed to coalesce around recommending an independent inquiry as the least offensive (and cheapest!) course of action. Even this, however, raised political issues of considerable delicacy. Any inquiry would have to be preceded by a meeting between ministers and the industry which, to be effective 'will have to include a blunt—and therefore bleak—exposé of the Government's assessment of the industry's future. If such a meeting was held in advance of an Election, would it not place a powerful weapon in the hands of your political opponents—would they not represent it as a confession of failure, after eight years of Conservative administration?'[76] This was hardly impartial Civil Service drafting but it did enough to frighten the Economic Policy Committee of the Cabinet into requesting yet another report and refusing to meet the Shipbuilding Conference. The attitude of the Treasury to even an independent enquiry now began to harden. The Secretary to the working party took the view that it was 'unlikely that an independent inquiry would bring forth to any extent facts, views or remedies', which the working party had not considered and that as such there was 'no real need for an independent inquiry'.[77] This attitude, in its turn, led to a full-scale attack on the core premises on which much work on the shipbuilding industry had been based since 1945. As one official observed:

> The main premise . . . is that the Government must take action to preserve shipbuilding capacity. This assumes (1) that demand on British shipyards is going to fall and then recover (2) that as a result of the decline in demand in the early 60s, so many yards will close that capacity will be insufficient when the market eventually improves (3) that it is better to preserve capacity during the lean years rather than let it go out of commission and be replaced with altogether new capacity if and when it is once again required (4) that activity must be maintained for the sake of a comparatively few craftsmen who might have difficulty in getting similar work elsewhere. All seem to me highly questionable.[78]

This stood all previous calculations on their heads in that it seemed to agree with the shipbuilders' analysis by accepting that the problem was too much capacity. In the Treasury view, if Government money was to be spent on shipbuilding then it should be in compensation payments, re-training grants, and 'encouragement to new industries to set up in the shipbuilding areas'.[79] This was, however, unlikely to ease the problems of the Government with regard to the industry.

The debate on the nature of the industry was about to become both highly public and controversial. Sir Graham Cunningham, the Chairman of the Shipbuilding Advisory Committee, had suggested on three separate occasions since mid-1959 the formation of a sub-committee to examine the problems of the industry. While the shipowning and union representatives had accepted the proposal, it had been rejected by the shipbuilders with 'excuses so frustrating that to continue serving the industry as independent Chairman would be fruitless'.[80] Moving quickly to limit the political damage, the Government installed the Permanent Secretary at the Ministry of Transport, Sir James Dunnett, as the new Chairman, and established a sub-committee of the SAC to report on the industry which, as *The Times* stated, 'justified' Cunningham's views.[81] Before the SAC report could be completed, however, another high-profile controversy embroiled both the Government and the industry, the leak of the Department of Scientific and Industrial Research report on research and development in the shipbuilding industry. This report had been with the industry for some time before *The Times* published a précis of its contents in October 1960. The industry's record on productivity and modernisation was woeful, production control in the industry was primitive, the total effort devoted to research and development was insufficient, and almost no organised research had been applied to production and management problems with a view to improving the productivity of capital and labour and reducing costs. Not surprisingly, this provoked a furious row and a rewriting of the report for publication which, as *The Times* commented, had 'been purged . . . of a great deal of serious and valid criticism contained in the original'.[82] All in all, this did little more than make the position of the Government worse. As the Minister of Transport, Ernest Marples, had observed to the Prime Minister, Harold Macmillan, in 1959:

> After twenty easy years it has only needed two bad ones to produce rumbling noises from the industry about Government help. I think they may be in serious trouble during the life of this Parliament. We must be ready in advance to cope with a crisis as a time of crisis is always a time of opportunity.[83]

Just exactly what this 'opportunity' was remained to be seen.

The recession of the 1950s certainly had a profound effect on the attitude of the Government which, whilst worried about the rising level of unemployment in the industry, quickly seemed to accept that some degree of contraction was inevitable.[84] The estimated position at the end of 1961 was that of 24 firms with 124 berths, there would only be 48 ships on the stocks and 17 to be laid down: 14 companies had no orders.[85] The Government, however, continued to dither over what action to take given that the industry itself seemed to be incapable of formulating a response to the situation. As the Board of Trade observed, there was little point in any Government action 'until the prospects of the industry are very patently bad'.[86] Not only did this reveal a thoroughly reactive stance but, given the situation, rather begged the question of what 'patently bad' actually meant? Certainly the increase in unemployment continued to worry the Government with total employment in the industry having fallen from nearly 81,000 in 1957 to 63,477 in 1961. Shipbuilding continued to show much higher rates of unemployment than the manufacturing industry average of 1.7 per cent in 1961; unemployment in shipbuilding ranged from 1.2 per cent at Barrow, 7.1 per cent on Clydeside and at Southampton, to 17.1 per cent at Belfast, with the national average being 5.2 per cent. While it was recognised that shipbuilding employment represented a comparatively small proportion of total employment, what concerned the Government was the prospect of heavily concentrated localised unemployment.[87] Employment in the industry, therefore, was falling, and this was exacerbated by the information that 'some shipbuilders were eliminating profits in order to secure orders', a practice redolent of the inter-war years.[88] Thus, while the situation may not have been 'patently bad', it was bad enough and was set to get worse, as the Government attempted to deal with a totally contradictory stance towards the situation on the part of the Shipbuilding Conference.

In public, the Shipbuilding Conference took a very bullish line on the situation. For example, the President of the Shipbuilding Conference rubbished those who criticised the low volume of export tonnage on the books and drew satisfaction from the fact that 'soundly based British shipowners . . . [were] less likely to cancel their orders than some overseas customers'.[89] This theme was taken up by the Managing Director of Lithgows, A.H. White, who again extolled Britain's conservatism and confidently predicted that 'the nations have overbuilt and some shipyards will go out of business . . . it will be a case of last in, first out, and that Britain's policy of modernising her shipyards without any great expansion of their capacity will pay dividends in her future competition with the "mushroom" yards'.[90] This attitude formed the basis for an editorial in the *Shipbuilding and Marine Engine Builder* in December 1960, which, whilst ostensibly concerned with 'diminishing order books', could still sound remarkably upbeat:

> the policy pursued by British shipbuilders is in contrast to that of the ambitious, younger maritime nations abroad who were encouraged by the boom years to establish shipbuilding industries or expand existing ones on a magnanimous scale. It could be that such 'mushroom' growth may be in danger of some blight if the icy economic winds continue to blow across the world of shipping. It could also be that the policy of consolidation British shipbuilding has followed will have been one of intelligent anticipation to a degree which was not perhaps acknowledged in all quarters.[91]

The same editorial noted that the tonnage being constructed overseas

Plate 5: The platers' shed at Stephen of Linthouse on the Clyde in 1950. It could just as well have been 1900.

for British owners—537,610 grt—was a new record and was double that on the stocks in the UK for foreign owners. It also advanced a novel explanation for this: the tonnage had been ordered when UK order books were full and 'while British owners are more universal in their outlook in the placing of orders, it is reasonable to assume that ordering abroad in such volume was abnormal and is not likely to be sustained as and when the flow of orders resumes a healthier trend'.[92] At every level this was wrong. By 1962, *Fairplay* was examining the second quarter returns from *Lloyd's Register* and concluded that 'Great Britain and Northern Ireland now has the smallest share of world construction in all her history, excluding war years'.[93] As yards began to announce closure plans or to close—Crown's at Sunderland, William Gray and Company at Hartlepool, the Wear Dockyard, Short Brothers at Sunderland and Mitchison's at Gateshead, all on the north-east coast; and Hay's at Kirkintilloch, the Ardrossan Dockyard, D. & W. Henderson, Denny's at Dumbarton, George Brown and Company at Greenock, Harland and Wolff at Govan, the Blythswood at Scotstoun, Simons and Lobnitz and A. & J. Inglis at Pointhouse—shipbuilders were united in the opinion that the Government had to act.

In private, by the third quarter of 1962 the Shipbuilding Conference was bemoaning 'the serious plight of the shipbuilding industry . . . and the depressing prospects for the future'.[94] The Conference calculated that the annual rate of completions had fallen by 200,000 grt between 1961 and 1962, with launchings and keels laid showing corresponding falls. Orders for the first nine months of 1962 represented a post-war low of only 317,000 grt whilst those for the third quarter amounted to 34,000 grt. The UK had taken only 7.6 per cent of total orders in 1962 as against West Germany and Japan, which took 12.7 and 30 per cent respectively. Whereas UK launchings had fallen in 1962, those of Japan, France, Norway and Sweden had increased and both West Germany and Japan had larger order books than the UK.[95] The Conference considered the situation bad enough to formulate a scheme for the reduction of capacity 'closely in line with . . . [that] in the inter-war period'.[96] The formulation of this scheme would eventually bring the Government to the realisation of just how bad things actually were. The problem for the Government lay in the nature of the scheme advanced by the Shipbuilding Conference to reduce capacity. This essentially re-envisaged the re-creation of the NSS of the 1930s. This body, which had been responsible for the closure of what had been deemed excessive capacity in the 1930s, but had incurred much hostility, provoking the Jarrow

March, and although the Government recognised the need for capacity reduction, it did not wish to be associated with such a proposal. The Conference scheme, however, was remarkably similar to that of the 1930s. It was proposed to use the Shipbuilding Corporation Ltd, a company established to run sterilised capacity during the war years and which had since been dormant, to utilise its reserve of £1.5 million in cash to buy up yards which wanted, voluntarily, to go out of business. Once the assets were used up, the industry would institute a levy and only then resort to the possibility of Government assistance. The Conference envisaged the liquidation of some 20–25 per cent of the industry.[97] The major obstacle to this was that in order to qualify as a 'redundancy company', so that the levy on the shipbuilders could be charged as business expenses and offset against income tax, the Shipbuilding Corporation Ltd, and the scheme, had to meet the conditions for such a body set out in the 1952 Income Tax Act.[98] These conditions stipulated that redundancies had to operate a strict plan concerning the units to be closed, the time-scale involved, and satisfactory redundancy compensation payments for displaced workers had to be agreed. The Conference scheme met none of these conditions, being voluntary, open-ended and leaving any details of compensation to individual yards.[99]

The Government was placed in an immediate quandary by the Conference proposals. As the Chancellor of the Exchequer, Reginald Maudling, noted, '[while] it was inevitable that the shipbuilding industry . . . should contract . . . the Shipbuilding Conference should be given no reason for thinking that their present scheme as it stood was likely to be acceptable'.[100] There were compelling reasons for the Government to reject the Conference scheme: given that no element of compulsion was envisaged, the scheme could not work to any pre-determined plan, and there was no definite target for capacity reduction nor satisfactory arrangements for redundancy. If the Government supported the Conference proposals, therefore, it would be 'implicated in the working of the scheme, although powerless to control it. It might be said that the Government rather than the slump in shipbuilding, were responsible for the contraction of the industry and the closing of particular yards.'[101] As the President of the Board of Trade, Frederick Errol, put it, the present plan 'would be bound to attract political odium'.[102] Rejecting the scheme, however, also carried a high degree of risk. The Government would then be open to the accusation that it was rejecting a 'policy for shipbuilding' and that this rejection in itself would be viewed as the core of the problem; nor would rejection of the scheme ease the pressure which was bound to fall on the Government to prevent the closure of particular yards. The Government could also be implicated because of 'the substantial public subsidies' received by the aviation industry and asked why shipbuilding did not qualify for similar treatment.[103] Just how laconic the Government's approach could be, however, is encapsulated in the following exchange between Macmillan and Maudling:

> PMs Personal Minute to the Chancellor of the Exchequer: 5 July 1963. The two great problems that confront us are cotton and shipbuilding. I should be grateful if we could have a talk about them.
>
> 8 July, PM, Thank you very much for your minute . . . I should very much welcome a talk about these two industries. I think I am fairly up-to-date with shipbuilding but I cannot claim the same about cotton. May I take a day or two to familiarise myself with the present position in the cotton industry and then come and see you?[104]

Given the weight and intensity, not to stress the longevity, of the briefing on shipbuilding one could only draw the conclusion that the situation in cotton must have been truly parlous. Ultimately, the Minister of Transport, Ernest Marples, drove the situation by arguing that the only way in which the Government could exert direct control via planned capacity reduction would be through payment of a subsidy. As Marples remarked 'we should not escape political difficulties by doing so, on the contrary—but we should have some control'.[105]

The alternatives canvassed were payment of selective subsidies to preserve certain yards, probably at the expense of others, or measures designed to increase the flow of orders. The Government therefore decided to seek modification of the Shipbuilding Conference proposals and utilise subsidies. The situation was given added urgency in that the Government seemed incapable of disassociating its own reading of the market from that of the shipbuilders. Despite the fact that an obvious recovery in international shipping and shipbuilding had begun in 1961, and that the trend of launches was rising, the collective prognosis for British shipbuilding remained pessimistic. It was considered, quite correctly, that few passenger ships would now be built, although this was viewed as an object of regret, given the degree of specialisation among British builders and the high percentage of added value which this type of work entailed. The trend towards larger cargo vessels, such as tankers and bulk carriers, was also an object of regret in that few British builders specialised in this type of market, the Wearside SD14 being a later and notable exception, and it also narrowed the market for the small and medium sized builders. On the best advice available to the Government:

> There is little prospect, apart from an emergency, of an early recovery; there is now a world surplus of shipping, and this is being aggravated by the building programmes still in hand . . . it would seem that capacity representing an output of some 0.4 mgrt of merchant ships (or roughly a quarter of total capacity) might disappear with little risk that the industry would be too small when demand recovers . . . unless there is a very marked upturn in the rate of ordering, the industry will be working at only one-third of its capacity by this time next year.[106]

The industry, therefore, had borne out the strictures advanced on the competitiveness of any individual firm or industry in a rising and falling market. The performance in the booming market of 1945–58 could only be described as disappointing. In the recession of 1958–61 it was disastrous. In just three years, merchant tonnage under construction in the UK had fallen by 45 per cent, there was no relief from the naval building, and employment had fallen by 43 per cent. With the Government estimating that employment in mercantile shipbuilding could fall from 60,000 to 45,000 between 1963 and 1965, there was a danger that the Government would be accused of preferring 'a chaotic free-for-all reduction of capacity', and that accordingly, whilst undesirable, 'it might be advisable to face up now to the need for subsidy'.[107]

The situation was remarkable. In the space of less than 20 years the British shipbuilding industry had gone from world leader to being little more than a 'basket case'. All of the familiar complaints had been aired, some by way of explanation, some by way of excoriation: the lack of decent credit facilities, late delivery dates, high prices and an insouciant disregard of new ship types. British builders, however, stuck to what they knew best in terms of ship types and building for trades and routes. As one of the Bank of England's industrial advisers saw it, 'the protective skin that tradition and convention gave to the home market has been split beyond repair'. This was a direct reference to the fact that British shipowners were, in increasing numbers, ordering their ships abroad. The sustaining base, therefore, of British shipbuilding, the home market, was increasingly going abroad. The Bank's adviser, A.C. Darby hammered the attitudes of British builders to layouts, plant, labour, credit and finance. Despite the fact that Darby believed, somewhat contradictorily, that the UK could compete on the world stage, he bemoaned the fact that credit restrictions, as well as sneering attitudes, had kept British builders out of the PanHonLib market, a conflation of the fleets of Panama, Honduras and Liberia, which were the fastest growing fleets in the world.[108] That the British ignored these fleets was at the least unfortunate, because they represented a step-shift in the world carrying trade and the types of vessels in which it would be carried. The stunning reversal of the industry's prospects would prove to be the harbinger of renewed Government interest in the future of the industry.

5

Things Begin to Slide

Throughout the 1950s Government was probably as well informed about the situation in the shipbuilding industry as it was concerning any other. The decade was marked by unofficial inquiry followed by unofficial inquiry in Government circles, inquiries by the Bank of England and a whole raft of quasi-official and non-official investigations. All produced mountains of evidence but little in the way of proposals around which the Government and industry could coalesce. All of the well-known ills were given an airing: the paucity of credit facilities, high prices, late delivery dates, technological stagnation, and the failure in marketing especially regarding new markets and new ship types—a failure which was particularly marked in the tanker and bulk carrying trades, especially the very large and ultra large crude carrier market. The British domestic market incurred substantial penetration, exports declined, and as yard closure followed yard closure, Government began to worry not only about the industry in balance of payments terms but also about the social costs of closure and the impact of heavily localised unemployment.

Given the plethora of information available to Government on the condition of the shipbuilding industry, inactivity on the part of the State may seem surprising. The problem for Government, however, lay in the contradictory nature of much of the advice and in establishing exactly the key point in breaking what increasingly appeared to be a vicious circle. While the Government could hardly shirk high-profile rows such as Sir Graham Cunningham's resignation and the leak of the Department of Scientific and Industrial Research (DSIR) report, just what to do with the welter of recommendations remained unclear. The DSIR report into the research and development requirements of the shipbuilding and marine engineering industries was scathing across the board. The industry's record on modernisation and productivity improvements was appalling, production control was primitive, and insufficient effort was devoted to research and development, with no organised research applied to production and management with the aim of improving the productivity of both capital and labour and thereby reducing costs. In the view of the authors of the report, British productivity was neither high enough nor its costs low enough to enable British shipbuilders to 'compete successfully with foreign shipbuilders'. The report commented that productivity in British shipbuilding had only risen by 1 per cent since the Second World War whilst costs had soared and capital investment lagged behind that of all other sectors of manufacturing industry. The most stinging criticism, however, was directed at the marine engineering industry and, in particular, the Parsons and Marine Engineering Turbine Research and Development Association. This organisation acted as the design and research development centre in marine engineering and was funded through a levy on the shipbuilding industry. It was criticised for its long-held commitment to steam turbines and slow-speed engines which, it was argued, had stultified the development of marine engineering in Britain, with only one indigenous firm, Doxford, producing its own patented diesel engines. This had given the larger continental companies—Burmeister & Wain, MAN and Sulzer—a competitive advantage which had led to British shipbuilders purchasing licences from European firms. The report concluded that in both shipbuilding and marine engineering the whole thrust of research and development had been directed at 'an efficient and reliable product', the finished ship. Shipbuilding had been slow to reorganise its production methods to exploit technological developments and none of the research organisations serving the industry had mounted investigations to improve and cheapen production methods.[1] The leak of the report to *The Times* and the furious row which then ensued, with the shipbuilders asserting that the report totally misrepresented the position of the industry, probably gave the Government pause for thought in terms of action.

The DSIR report was followed by that of the SAC Sub-Committee which, of course, had only been established in the wake of Cunningham's high-profile resignation. Given the circumstances surrounding this report, effectively forced into it by the row, and noting its constituency—shipbuilders, shipowners, trade unions and officials—it was hardly surprising that it failed to reach agreement on many points. Its nine major recommendations were hedged around by qualifying language—'consider, bear in mind, review'—and full agreement was reached on only one point, the need for Government assistance via a credit scheme. To the Sub-Committee this was the only acceptable form of aid in that it obviated any prospect of Government intervention in individual yards.[2] In an attempt to cut through the contradictions of official or quasi-official advice, the Government decided to commission some independent research as to why British orders were going to foreign shipyards and duly asked Peat, Marwick, Mitchell to report. The consultants found that the main reasons for orders being placed abroad were: price; price and delivery date; price and credit facilities; guaranteed delivery dates; and the reluctance of British builders to install foreign-built engines. Its main conclusion was that 'the availability of credit facilities to spread payment for a ship over a number of years does not appear in most cases to have been of primary importance'.[3] Thus the central recommendation of the SAC report was rejected by Peat, Marwick and Mitchell, and, all the more discommoding for Government, produced a totally contradictory stance on an issue—credit—which had rumbled back and forth between Government and the shipbuilding industry since the Second World War.

The issue of credit for shipbuilding had first emerged during the inter-war years as a consequence of the rise in costs which had occurred in the war and immediate post-war years. The standard contract tended to involve the payment in stages by the shipowner to the shipbuilder over the life-cycle of the construction of the vessel, the whole ship being built on a cost-plus basis, with ultimately the shipbuilders' bankers acting as the financier of the whole transaction. Whilst this arrangement lasted in some form or another until well into the 1960s, it did not survive in its pure form after the First World War. In 1924, for example, the Cabinet Committee on Unemployment considered, although ultimately rejected, a proposal that Government should intervene by advancing 50 per cent of the cost of new tramp steamers, interest free for the first five years and then bearing 2.5 per cent for the next five.[4] Shipbuilding made more use of the Trade Facilities Act than any other industry by taking over £20 million of the £73 million guaranteed.[5] More Government assistance was forthcoming with the underwriting of the insurance on the Cunard liner *Queen Mary* under the 1930 Cunard Insurance Act, the rescue of the Royal Mail Steam Packet Company and further assistance through the British Shipping (Assistance) Act of 1935. Even in the inter-war years, therefore, there was some recognition that the traditional method of financing ship construction was already under strain. After 1945, with further rises in all costs, the issue took on even more significance. *Fairplay*, for example, had regularly monitored the cost of a 9,500 dwt vessel of 12 knots and recalculated the price index to a base of 100 for January 1945, and revealed that the index had risen to 162 by January 1950.[6]

The credit issue became immediately problematic after the war when a combination of dislocated markets and increased costs brought it to prominence. This first manifested itself in negotiations with Norway, traditionally Britain's most important export market, where, after some years of wrangling, the British Consul in Oslo eventually informed the Foreign Office and Treasury that a failure to ease credit terms might lead Norwegian owners to 'divert substantial orders to other countries . . . where better credit terms are available'. Suitably provoked, the Foreign Exchange Control Committee of the Treasury and the Bank of England met in emergency session and, given 'the desire not to lose our best foreign customer to Swedish and other yards', they decided to 'grant to Norwegian shipowners credit terms of fifty per cent on completion and fifty per cent over five years'.[7] This, though, was little more than an indication of trends and, inspired by the creation of a number of specialised continental European institutions that made loans for financing ships, interested parties established such a company in Britain. The Ship Mortgage Finance Company (SMFC) was founded in March 1951 under the chairmanship of Sir James Milne, the head of Coast Lines. It was formed as a private company, with an authorised share capital of £1 million, divided into 1 million ordinary shares of £1 each, on which initially 1s was paid, and borrowing powers of £10 million. Three institutions subscribed over half the capital: the shipbuilders trade association—the Shipbuilding Conference—acquired 0.25 million shares through Birchin Lane Nominees; the Industrial and Commercial Finance Corporation subscribed for 0.2 million; and Hambros Bank, a traditional player in London ship-market financing, took a further 0.1 million. The remainder of the shares were taken up by the leading insurance companies.[8] Despite what appeared to be a positive move, the creation of the SMFC was bitterly opposed and indeed resisted by the Bank of England on the grounds of exchange control and fears of setting off a credit war. These reasons, coupled with consistent credit squeezes throughout the 1950s, meant that over fourteen years the SMFC assisted in the financing of some 60 ships by providing £16.5 million from its own funds and funding participants for a further £5 million. The history of the SMFC, therefore, even from the pen of its own Chairman is one of under-capitalisation, missed opportunities and counter-factual regret.[9] Although the shipowners and shipbuilders had periodically complained to the Government over credit terms, both the scale and longevity being criticised, it was the impact of the credit squeeze in 1957 which highlighted the problem.

Throughout the 1950s the issue of credit acquired ever-increasing importance. In Britain this had little to do with the usual vagaries of the shipping freight market and all to do with the underlying economy. The systemic weakness of sterling and the balance of payments constantly combined to frustrate the achievement of industrial and economic stability. The various attempts to stabilise the economy made by the 1945–51 Labour Governments were all but ruined by the impact of Korean War rearmament and from 1951 the economy was subjected to successive bouts of 'stop–go' economic policies, the underlying ambition of which was to hold the exchange rate at $2.80 in order to prepare for full sterling convertibility. As restrictions on sterling convertibility were progressively removed over the 1950s, sterling was afflicted by regular crises of confidence. Accordingly counter-measures were taken, the general impact of which was to dampen demand in the domestic economy. In 1951, for example, the Labour Chancellor, Hugh Gaitskell, withdrew the 40 per cent investment allowance which his predecessor, Sir Stafford Cripps, had instituted in 1949 to stimulate industrial re-equipment. Capital issues and bank advances were restrained, import restrictions were imposed, excess profits tax were reintroduced and there was the first post-war increase in the Bank Rate. In the following year there was a full-blown balance of payments crisis which involved further import restrictions, capital investment cuts, hire purchase restrictions and another rise in the Bank Rate to 4 per cent. Although policy was relaxed in the course of 1953–54—indeed the Chancellor, R.A. Butler, went so far as to reintroduce the investment allowance—by the end of 1954 a decline in the free market price for Transferable Account sterling and heavy sales against the dollar led inexorably to a tightening of policy. In January and February 1955 the Bank Rate was raised twice and the financial authorities were compelled to intervene in the foreign exchange markets.

Further measures to restrict bank credit and capital investment were taken in the summer and autumn of 1955, and by 1956 the Bank Rate was again raised to 5.5 per cent and investment allowances once more abolished. The impact of the Suez crisis not only forced further restrictions in lending but dealt a major blow to international confidence in sterling. The overall impact 'was to restrict severely the use of sterling as a currency for the financing of international ship acquisitions, i.e., for pretty well all transactions other than new buildings'.[10] The outcome was to make the dollar and the eurodollar the major currencies in international shipping finance, and whilst the few specialist finance houses in this market, notably Hambros, moved to build up their dollar base, the general situation left the normal channels of British credit in a weak position. The issue came to a head when the Chancellor of the Exchequer, Peter Thorneycroft, introduced a package of proto-monetarist measures. One commentator termed this 'the sudden violence of the 7 per cent bank rate in the autumn of 1957', and went on to claim that:

> It was not only that the halt to the further plans of the nationalised industries made it certain that the growth of industrial investment, which had been proceeding steadily since 1953, would now be reversed. The Government this time succeeded in creating an atmosphere of industrial anxiety, and this went a long way to removing any lingering enthusiasm for expansion in private manufacturing industry. Everyone suddenly felt the wisdom of postponement.[11]

Predictably enough, the shipbuilders, reluctant to spend their own money throughout the 1950s, were keen to spend taxpayers' money through improved credit loans.

As the General Council of British Shipping had observed in a memorandum to the Committee on the Working of the Monetary System in 1957, 'until then the money market in general had responded reasonably well to the needs of the shipping industry in that the gap between the short-term finance traditionally offered by the banks and the long-term loans available from the financial institutions was being satisfactorily closed'.[12] The shipbuilders, however, remained to be convinced. Responding, for example, to the 1951 credit restrictions, which stipulated that bank credit for overseas purchases of capital goods had to be repaid within six months, the Harland and Wolff Chairman, Sir Frederick Rebbeck, was caustic: 'Consider the position. The owner takes his ship and tries to earn enough money in half a year to pay off that debt. It is too ridiculous for words.'[13] In 1953 the SMFC Chairman, Sir James Milne, highlighted the 'heavy decline in orders placed for construction of new ships in British yards'. He further remarked that with all countries in Western Europe and the USA having either institutions or Government arrangements supporting shipbuilding and shipping through cheap credit, if SMFC were not given wider latitude in the markets, then the results would be the 'undermining [of] the stability and competitive power of both the shipping and shipbuilding industries'.[14] The Shipbuilding Conference's 1953 annual general meeting considered figures for tonnage commenced, launched and completed between June 1952 and June 1953. It noted that the UK had registered a decrease under every heading, whereas the rest of the world 'showed a very substantial increase under each head'.[15]

The contradictory policy stance of the Government was also exposed when the Ministry of Transport contacted SMFC, informing it that the British Government had been informed by the Norwegian Government 'that unless more finance was made available in the UK for Norwegian shipping a substantial number of existing orders for Norwegian shipowners for ship construction in this country would not take place'. Asked by the Ministry as to what SMFC could do to alleviate this position, the somewhat doleful reply was that with the Bank of England having limited lending to foreign shipowners, restricted the company's borrowings and forced it to raise long-term capital through debenture issues, the answer was very little indeed.[16] The response of the Shipbuilding Conference to this state of affairs was one of ill-disguised fury, with a demand that the Treasury remove restrictions on the provision of credit facilities as 'the lack of unrestricted credit facilities is a serious handicap to British shipowners wishing to build new ships in British shipyards'. Describing the assistance from the SMFC as 'limited and restricted', which it was, the Conference argued that foreign shipowners seeking credit facilities were turning to other countries where 'longer term credits at low rates of interest are being made readily available by banks or by special financial institutions under government influence'. The Conference went so far as to suggest the provision by the Treasury and the Bank of England, working through the SMFC, of a specialist institution to provide credit facilities for ship construction.[17] Such a proposal was, however, unlikely to move Government since, despite the Shipbuilding Conference's attempts to marshal the necessary evidence, there was little indication that credit facilities alone, as opposed to issues such as prices and delivery dates, were the main reasons for orders being lost.

As the General Council of British Shipping noted, the impact of the 1957 credit squeeze was that 'the possibilities of medium-term loan finance have therefore virtually disappeared for the time being'.[18] As the SMFC annual general meeting in 1960 stated:

> The shipbuilding industry is facing a critical period, and it is very desirable that no condition important for its competitive success should be lacking. As against this situation, there is a stringent paucity in this country of medium term finance available for lending on mortgage to shipowners.[19]

Credit then had been a consistent irritant throughout the 1950s. None of this, however, goes any way to explain the Government's response to the raft of reports in the early 1960s. The DSIR report took no notice of credit, its brief being to examine the research and development requirements of the industry. The single issue on which the members of the SAC report could agree was the need for credit facilities, although the Government could easily dismiss this as special pleading from an interested party. The Government's own 'independent inquiry', that of Peat, Marwick, Mitchell, specifically rejected credit terms alone as being responsible for British orders going abroad, all of which left Government in the bind of having received totally contradictory advice on one of the central issues facing both it and the industry. The Government was caught in a quandary. Shipbuilding unemployment stood well above the all-industry average.

The potential redundancy scheme, brought forward by the Shipbuilding Conference, raised all of the spectres of the inter-war period with the possible political odium which would then ensue. Furthermore, the onset of yard closures—five in 1962 with another eight announcing their intentions to close in 1963—in all probability concentrated the Government's mind in that facing up to subsidy was preferable to a chaotic free-for-all. With discussions proceeding with the Shipbuilding Conference on a modified redundancy scheme, and within the Government on the form of the subsidy, the Government was jolted when the Conference rejected the modified scheme for redundancies, with the shipbuilders refusing to accept the conditions concerning planned redundancies and compensation. Rejection of the Government's scheme on the one hand did not mean that assistance would be rejected and the Conference pleaded for 'an early positive statement . . . of the Government's intention to stimulate demand for new ships from British owners, and of the industry's objections to direct subsidy'. The direct subsidy of shipbuilding, it was considered, 'would represent a dangerous step on the slippery slope towards nationalisation'.[20] If the Government feared that its rejection of the Conference scheme would increase the demand for subsidy, the Conference rejection of the

Government's position effectively achieved exactly the same result. With the Government considering that orders would recover from the poor level of 1960–62—it still believed that 'significant surplus capacity' would remain and that 'permanent contraction . . . [was] inevitable'—it was decided that announcing a credit scheme was the least worst option.[21]

The Shipbuilding Credit Scheme, announced in May 1963 and to run for one year, provided £30 million in funds, at the Government lending rate, for 80 per cent of the cost of a ship and loans which could be up to ten years. The financial limit was twice extended in the course of the year, eventually totalling £75 million. By October 1964, the scheme was fully subscribed with 67 vessels totalling 892,000 grt on order. Table 24 provides a breakdown of both the regional impact of the scheme and the type of vessels ordered.

In a sense the introduction of the Shipbuilding Credit Scheme contradicted the evidence available to the government—most notably from the Peat, Marwick, Mitchell report—that credit facilities were not an important factor in the dearth of orders by British owners being placed in British yards. The available evidence to the contrary, however, had not stemmed the tide of complaint, with the SAC able to comment in 1963 that 'the level of ordering by British owners has fallen off sharply in recent years'.[22] The origin of this scheme still remains difficult to discern, but the Government does seem to have been concerned by the actual and potential scale of unemployment and to have felt somewhat boxed-in by the range of options, especially close to a general election. If, however, assessing the origin of the schemes is difficult, calculating its impact is even more difficult given the problem identified by one commentator of 'establishing what would have occurred in its absence'.[23] The standard conclusion probably remains the best, in that the impact of the scheme was probably to accelerate orders which would have materialised in any case, although it did serve to concentrate orders 'in a period when they were much needed'.[24]

If credit was a problem shared by the industry and the Government then another post-war difficulty, marketing, or rather the lack of it, resided wholly in the attitude of the shipbuilders to the market place. Although building for export was not a particular British specialism, it did represent an important element in the order book normally accounting for an average of 25 per cent of output. In historical terms, Britain's most lucrative overseas market was Norway. In the nineteenth century Norway was an important market for the sale of second-hand

Table 24: The Shipbuilding Credit Scheme: type, number and gross tonnage of ships

	Builder Undecided		Scotland & Wales		N. England		Rest of England		N. Ireland		Total	
	No.	gt	No.	gt	No.	gt	No.	gt	No.	gt	No.	gt
Passenger vessels	1	58,000					1	116			2	58,116
Cargo vessels over 15 kts	1	6,500	5	50,220	8	71,393					14	128,093
Bulk carriers			10	181,418	13	299,900			2	57,000	25	538,318
Tankers			4	51,420	3	66,200			1	37,000	8	154,620
Coasters					4	6,398					4	6,398
Misc.	1	500	2	2,900	9	1,620					12	5,020
Non-self-propelled			1	1,750			1	145			2	1,895
Total	3	65,000	22	287,708	37	445,491	2	261	3	94,000	67	892,960

Source: PRO, BT199/14, Shipbuilding Advisory Committee, February 1964.

vessels but it became a major customer for new ships in the twentieth century. In the decade before 1939, the Norwegian maritime marine added some 2 mgrt of foreign-built ships to its register. Britain had contributed 440,000 grt of this, or 22 per cent of the total. There were good reasons for believing that Norway would be an important post-war market, and even before the war had ended the Norwegian broking firm A/S Securitas Ltd wrote to a number of British shipbuilders predicting the future level of demand. With the Norwegian mercantile marine ravaged by wartime losses, and the country having a small-scale shipbuilding sector, 'therefore in the coming years large building orders for the Norwegian account will be placed with foreign shipyards'. As A/S Securitas was attempting to sell its brokering services it was hardly surprising that the firm pointed out that British tonnage for Norway had fallen from 65 per cent of the total in 1930 to only 10 per cent in 1938, a decline which the firm ascribed to the lack of representation in Norway.[25] Such was the importance of the mercantile marine to Norway's balance of payments, which was, as the British consul commented, 'out of all proportion to Norway's population', that the Norwegian Government had made the restoration of the fleet a national priority.[26] Indeed, the early battles over the issue of credit were conducted with regard to the Norwegian market and by the middle of 1949 the British Government had gone so far as to grant Norway 'most favoured nation treatment in the matter of shipbuilding credits'.[27] Such a strategy appeared to meet with success. As the Norwegian mercantile fleet almost doubled in tonnage terms, from just over 4 mgrt to 8 mgrt, between 1948 and 1956, British shipbuilders took a significant share, as Table 25 illustrates. Between 1948 and 1956 ships for the Norwegian account constituted some 30 per cent of all British launchings for overseas registration and Norway was the industry's most important foreign customer.

As the Norwegian and indeed world market for ships boomed, the reluctance of British builders to increase capacity was reflected in a loss of market share; nowhere was this more acutely reflected than in the Norwegian market. In the wake of the Suez and credit crises, Britain's share of the Norwegian market all but collapsed. Table 26 demonstrates the decline. All of Britain's major competitors in the Norwegian market fared better. Japan's output for Norway rose from 13,000 grt in 1956 to 331,000 grt in 1965, Sweden's grew from 288,000 grt to 772,000 grt, West Germany's from 175,000 grt to 298,000 grt, whilst Norway's own output increased from 167,000 grt to 329,000 grt, between the same

Tables 25: Vessels launched for Norway in the UK, 1948–1956

Year	No. of ships	GRT	% UK share of total Norwegian tonnage
1948	37	133,474	31.6
1949	39	255,864	44.0
1950	20	139,212	27.8
1951	30	308,787	48.4
1952	17	151,699	32.4
1953	18	135,992	23.4
1954	22	237,334	31.1
1955	19	234,043	30.3
1956	12	144,114	16.4

Source: *Lloyd's Register of Shipping.*

Table 26: Vessels launched for Norway in the UK, 1957–1965

Year	No. of ships	GRT	% UK share of total Norwegian tonnage
1957	2	18,968	2.1
1958	10	113,045	8.5
1959	3	28,270	2.7
1960	5	71,526	8.1
1961	8	121,873	10.5
1962	7	94,994	7.2
1963	6	204,500	13.8
1964	0	0	0.0
1965	1	51,972	2.8

Source: *Lloyd's Register of Shipping.*

dates. In other words, those shipbuilding industries which had modernised, re-tooled and re-oriented themselves in modern world markets had hammered British sales in Norway, and even more worryingly were beginning to penetrate the British domestic market.[28] Britain's share of the Norwegian market, which had averaged 30 per cent between 1948 and 1956, collapsed after 1956 to an average of just over 6 per cent between 1957 and 1965. The scale of the collapse is dramatically illustrated by the fact that in every year, with the exceptions of 1957 and 1964, Norway remained Britain's most important export market. Britain's performance in the 1956–67 period was all the poorer in that the Norwegian mercantile marine expanded from 8 mgrt to over 15.5 mgrt. The situation, however, merely confused the shipbuilders, as Mr Rae of the Shipbuilding Conference reported following a tour of Scandinavia:

> We say that one of the headaches for the shipbuilding industry in the future is likely to be the very considerable excess world capacity in relation to any foreseeable future demand. In my recent visit to Denmark, Sweden and Norway I found not only an apparent lack of concern about this but a good deal of what one might call expansion mindedness. New berths, new shipyards, as well as modernizations designed to increase output, were all current talk and seemed to be allied to an almost naive faith in the likelihood of substantial demand in future—and for big ships.[29]

Just exactly who was being 'naive' in this regard, however, remained a moot point.

Once again, reports and inquiries underpinned any consideration of these difficulties. By the early 1960s the employers organisation, the Shipbuilders and Repairers National Association (SRNA), had established British Shipbuilding Exports Association (BSEA) in a belated attempt to arm itself with a marketing arm. One of its first acts was to comment that Norway represented 'the second most important and steadiest growth market for ships in the world'.[30] The report went into considerable analytical detail regarding the Norwegian market but it also advanced stiff criticisms of British shipbuilders. As the Director of SNRA, R.B. Shepheard, commented, the report disclosed:

> a widely held, disturbing and, for us, a very damaging state of opinion of the British Shipbuilding Industry among potential

> Norwegian customers . . . they have voiced a series of complaints, the principal of which . . . is that in addition to regarding us as unreliable over delivery, many are now saying that we cannot be trusted to honour contracts.[31]

In fact, the report was caustic regarding the attitudes of British shipbuilders, who, despite having built for the Norwegian market over many years, and with Norway as their largest and most important overseas market, understood neither the market nor the Norwegian shipowner. As the report made clear, shipowning was the 'most important single factor in the Norwegian economy . . . [and] because of this the fortune and experience of one owner automatically becomes the concern of all the others'.[32] Any problems, therefore, with one contract in one British shipyard automatically became the preserve of all Norwegian shipowners, with all that this entailed. British builders tended to treat Norwegian shipowners in the same manner that they did British owners. Such an approach, however, failed to recognise two key differences. British owners were predominantly operators of sophisticated liner tonnage working within conferences. On the other hand, the staple fare of the Norwegian industry was tramping. Allied to this was the fact that Britain was a major importing and exporting economy whereas Norway was not. The British mercantile marine was largely tied to British internal and external trade while that of Norway was mostly concerned with trade outside of Norway and, as a consequence, was exposed to world-wide economic trading factors. The Norwegian owner, therefore, was a sophisticated tramp operator functioning in a world market with all of the stress and strains that this imposed. This global operation led to a form of business which British shipbuilders commonly misunderstood, although they had been given ample warning, in various reports, of the relationship between shipowners and brokers. As the BSEA report made clear:

> No shipowner in Norway operates without the services of brokers. Selling ships therefore to Norwegian owners requires a close relationship with and knowledge of Norwegian shipbrokers. The Norwegian broker combines the activity of a freight and shipbroker. His activities and connections are world-wide. He arranges the freight contract, nurses the owner and the builder; establishes contact with the user of the space, and is the vital link in the purchase of a new ship.[33]

This relationship placed a premium on delivery dates as the new vessel was often tied to a time charter and thus failure to deliver on time was a particularly damning indictment.

As the report commented, the British shipbuilding industry had two inbuilt advantages with respect to the Norwegian market: the British were well-liked and the Norwegians would rather build in Britain 'provided we are competitive'. However, 'the disadvantage of this . . . lies in the fact that rightly or wrongly the damage done to our image is all the greater when we default'. Thus:

> Our first task should be to do everything possible to meet reasonable Norwegian demands for damages on current contracts which have gone sour. Some of the owners concerned have lost badly on charters fixed at the time of concluding the contracts. This is of course the way much of Norway's shipping works. The Japanese, the Danes, the Swedes, and West Germans know this and recognise it—they meet it by delivering on time. We have not delivered on time and so should meet it by as flexible and generous an approach as possible. Such an attitude will do more than anything else at this time to improve our image.[34]

Over time, however, the image of British shipbuilders had taken a battering. The author of the report, Stopford Holt, commented that he had been asked 'time and time again' why the British went to such apparent lengths to 'emphasize . . . woes and tribulations and . . . inability to cope effectively with these troubles'. Holt claimed that every reported speech by a British shipbuilder in the Norwegian press usually comprised a list of excuses for poor performance, ranging from official and unofficial stoppages, shortages of labour, failings on the part of subcontractors, modernisation schemes not producing the anticipated results, to recently completed contracts having entailed substantial losses. The impression thus gained by the Norwegians, according to Holt, was of an industry where the shipbuilders had no control or responsibility over problems, and worse, had no ideas as to how to address the problems.

In a wide-ranging survey of the Norwegian market, Holt was simultaneously optimistic and pessimistic over Britain's future prospects. He went into painstaking details as to the volume of reference material on shipping available in Norway, usually published by brokers, and suggested that at the most basic level British shipbuilders should acquaint themselves with this material and make connections with brokers, otherwise they stood little chance of gaining orders. The report highlighted the importance of tankers in the Norwegian market but stressed that ordering would be 'for very large ships'. In terms of combined bulk carriers, the Norwegians had 'been in the forefront of this development' and interest in them had been 'pronounced'. The Norwegians, however, had 'established satisfactory relationships . . . with Swedish and Japanese yards and wonder why British shipyards have devoted so little attention to them'. In the liner trades, Holt forecast that this market would be altered beyond all recognition by the 'container revolution' and predicted two likely trends as a consequence: first, that owners would combine to raise finance to enter the container market, and second, that liner operators were likely to switch into other trades in all probability involving specialist ships. Holt also provided details on individual owners, noting that Wilhelm Wilhelsen was at an advanced stage with a design for a container vessel, but that he held 'such strong views on British shipbuilding . . . [that] to get him to order British is going to require a major selling effort'. Olsen and Larsen, on the other hand, were probably 'worth talking to' as neither held 'such very strong views against British shipbuilders'. Holt concluded his survey, somewhat plaintively, by observing that British yards should immediately and urgently address their problems in the Norwegian market. His results and suggestions were the distillation of many months of research and interviews with over 150 Norwegian shipowners whose views had been given 'with a genuine desire to assist the industry to regain its position as Norway's natural supplier of ships'.[35] Despite the thoroughgoing nature of the report it was too little and too late. Britain's hold on the Norwegian market was decisively broken in the 1950s and increased competition from shipbuilding industries with radically different outlooks in terms of production and marketing techniques, combined with Britain's lack of response in such areas, ensured that there could be no way back into the Norwegian market once it had been lost. In its most important and rapidly expanding market, therefore, the rise of international competition, and Britain's failure to meet it, led to comparative and competitive failure.

The lacunae in marketing, however, went wider.[36] As the report of the Shipbuilding Inquiry Committee (SIC) noted in 1966, 'the industry's attitude to the various aspects of marketing is still largely traditional'.[37] This was hardly surprising given that a near century long domination of shipping and shipbuilding markets by the British had inculcated attitudes which were difficult to break. The dominance of the UK mercantile marine in the carrying trades of the world had bestowed upon British shipbuilders an easy dominance in world markets. A hugely bespoke market had been built up between owners and builders, and a massive range of such relationships had been established such as those between the White Star group and Harland and Wolff, Cunard and John Brown's, and the Greenock Dockyard and the Cayzer group, which actually owned the yard. Very few yards, however, did not have client–builder relationships which had evolved over many years and these formed the basis of what was, by any definition, a substantial preferential market. Marketing, therefore, was a loose association of historical builder–customer associations, heavily based on a personalised relationship between the chairmen of the relevant companies.

Such relationships, in turn, conditioned the approach to building; it was to order, the customer was always right and the shipyards were required to maintain the widest possible spread in terms of ship construction. The ordering pattern, therefore, was dominated by the shipowners, being dependent on enquiries for the range and type of vessels built, and was entirely based on the relationships which existed at the level of the board or chairmen. It was, as one historian has noted, 'a passive marketing strategy, in which the shipbuilder is production rather than market-oriented'.[38] Such attitudes were well ingrained, one could almost say historically conditioned, and it was unsurprising that British shipbuilders were reluctant to change their attitudes and approach. In response to a welter of evidence on the situation, the shipbuilders took the view that any advice verged on the 'smart-arsed' and failed to understand the historical situation of British shipbuilding. Most of the advantages which British builders had enjoyed, however, had either disappeared or were fast disappearing. Riveting had been eclipsed by welding and the coal-fuelled steam engine had been superseded by the oil-powered diesel. The age of the great liner was passing in the face of emigration restrictions and the challenge from increasingly high volume and lower cost aviation. As one commentator noted, the numbers crossing the Atlantic by sea did grow by 20 per cent between 1952 and 1956 to exceed 1 million in the latter year, but in the same period the number of air passengers trebled to reach 1.2 million.[39] Although travelling by sea remained cheaper than by air, the margin was falling steadily. The alternative liner market of cruising appeared to offer little more than a market which would be the preserve of the rich. The ship market was also changing radically. The innovation of the very large crude carrier (VLCC), followed by the ultra large crude carrier (ULCC), revolutionised yard layouts and building techniques, with many European and Japanese shipyards abandoning traditional building patterns and adopting techniques based on production engineering. The trend towards ever larger, standardised crude carriers marginalised two former British strengths, custom design and a high value added level of fitting out. Nor did a former strength, naval building, plug any gaps, as the naval programmes remained emasculated by pressures on the defence budget and strategic uncertainty. The British mercantile marine no longer reigned supreme on the trade routes of the world, with the rise of the Greek, Norwegian, Japanese and open registry fleets—notably Liberia, but also Panama and Honduras—all mounting effective challenges to British shipping. Trades had changed,as had routes, and the decline of the British Empire only served to increase the strain as newly independent countries strove to achieve economic as well as political independence. With world orders rising remorselessly after 1960, all of these changes seemed to pass British shipbuilders by.

The Shipbuilding Conference did, however, attempt to rectify this paucity in its market stance with the appointment of an executive in charge of exports in 1962, a move which was codified in 1964 by the establishment of the British Shipbuilding Exports Association Ltd. BSEA had a broad mandate: collecting overseas intelligence, advising on relative credit terms, and negotiating export orders. By 1967 it had started to produce intelligence reports on overseas markets, trends in major fleet flags and in individual commodity trades.[40] By the mid to late 1960s BSEA had produced a mass of evidence which was enhanced by the formation, in 1966, of the Market Research Section of BSEA. Having spent years existing in an information-free environment, British shipbuilders were suddenly deluged with market, trade, commodity, credit and vessel design information. The question remained, however, could they utilise it? As one historian has observed, the 'shipbuilders found that they could make little use of the information since they lacked their own market research units which could have related the general information to the interests of the specific firm'.[41] But such views hardly reflected reality. Austin and Pickersgill had a huge success with the standardised SD14 bulker and other yards had some success in very specialised markets, with Hawthorn Leslie, for example, attempting to break into the LNG/LPG (liquid natural gas and liquid petroleum gas carriers) market. Furthermore, the reports were not so specialised as to be useless, as Stopford Holt's report on Norway correctly forecast that the container ship would be a major growth area.

Plate 6: A frame-bending squad at Lithgows *c.*1957, a photograph which could well have been taken in 1900.

The reports themselves were not the problem; rather, this lay in the attitudes of British shipbuilders to them. Such attitudes were long-held. Appearing before the North East Coast Institute of Engineers and Shipbuilders in 1947, F.A.J. Hodges presented a paper entitled 'The Application of Modern Management Methods to the Shipbuilding Industry'. Hodges's premise was that compared with other manufacturing industries and shipbuilding overseas, British shipbuilding had become relatively outdated and stultified by traditional methods. Hodges provided idealised shipyard layouts to cater for the full development of flow line production, including prefabrication and enhanced loft work. He suggested that most companies could reduce the number of berths and thereby enhance the space available for sub-assembly work, modern roadways and cranage to supply the remaining berths, and through a reduction in berth/erection cycles increase the throughput of steel and the output of ships. Yard modernisation, Hodges argued, could not be accomplished in piecemeal fashion but had to be part of a meticulously planned exercise covering every aspect of the shipyard. All of this, however, only served to raise in the minds of the shipbuilders their great *bête noire*, the prospect of overcapitalisation. Even more worryingly for the shipbuilders, the essential concomitant of new technologies and new yard layouts was new forms of management. The traditional system, therefore, of leaving the production process to the shipyard manager and the foreman was, Hodges suggested, inadequate to the orientation of the modern yard. What was required, it was suggested, was a Planning Department, a Rate Fixing Office, and a Production Control Department, and considerable detail was appended as to how each section would work and how they fitted into modernisation schemes. Fully aware of the conservative nature of the audience to which his paper was addressed, Hodges entered a final plea on behalf of new technology and new management methods: 'Undoubtedly, eventually they will be adopted either willingly, by an earnest endeavour to secure a thriving industry or, reluctantly, forced on us by the inexorable pressure of world competition . . . Modern management has only one purpose which is: to raise productivity to the highest possible level and to maintain it in that position.'[42] This was a prescient observation, but the fact that the same criticisms would still be made in the 1960s and 1970s was, to say the least, depressing. The reaction to Hodges's paper, however, would set a long-running tone with regard to modernisation, however defined, in British shipbuilding.

The discussion following Hodges's paper revealed the opposition to any forms of modernisation. Indeed, altogether new reasons were advanced for the inter-war slump: 'efficiency experts', it was claimed, 'are so very clever that after the end of the 1914–18 war they succeeded in some cases in persuading some shipbuilders to let them attempt reorganizing their works and when they had finished the job the yards closed up'. The planning, rate-fixing and production departments were variously viewed as useless, unnecessary, interfering with established practice, very costly or simply 'soulless'. Jim Lenaghan of Fairfield's went so far as to argue that 'any reorganisation which tends to robotize the foremen and take initiative from them, is not in the interests of production'. Even some of the positive comments were worrying, with Reg Ibison, the Hawthorn Leslie Managing Director, commenting that material on shipbuilding modernisation was 'extremely scarce', and that the industry was 'witnessing a revolution so wide in its application and so broad in its effect as to be of paramount importance to all engaged in the shipbuilding industry'. Such an indication gives some credence to the view that the 'revolutionary' methods of building in USA had been widely ignored. A somewhat surprising response to Hodges's paper came from the Harland and Wolff Chairman, Sir Frederick Rebbeck, who found the paper 'excellent', open for neither argument nor discussion 'but rather for complete agreement'.[43] This seems a somewhat surprising stance given the comments of the historians of Harland and Wolff regarding the launch of the *Canberra* in 1960:

> her hull had riveted frames and three riveted seams; she was, indeed, the last large vessel built to incorporate significant amounts of riveting. The use of riveting reflected the view which had been held in Belfast and by some owners that riveted ships were superior to welded vessels, a belief which delayed total commitment to welding at a time when emerging shipbuilding countries, such as Japan, Sweden and Germany, were producing all welded vessels. Another factor preventing the adoption of welding at Belfast was the inadequacy of cranage, which dated from wartime reconstruction. Sir Frederick discouraged contact with shipbuilders in Britain or abroad, and new initiatives from within the company were likely to meet with rebuffs . . . there were people in the Company who saw the need for change, but when reasonable profits were being made using largely traditional methods, incentives to modernise were simply not strong enough.[44]

This says it all: reasonable profits with traditional methods, and, therefore, no incentive to change.

The denouement came some thirteen years later in exactly the same forum in which Hodges had delivered his paper: a symposium on the reorganisation of some Tyneside shipyards. The Chairman of the symposium, G.H. Houlden, argued that the change from riveted to welded construction presented the shipyards 'with their most fascinating reorganization problem. . . since the change from wood to iron shipbuilding'. W.P. Heckles of John Redhead and Sons Ltd acknowledged Hodges's concluding warning by conceding that the company had recognised from the 1930s the demand for vessels of greater length, finer coefficient and increased deadweights, but that it had been the development of old and new competition which had 'necessitated' modernisation. Swan Hunter and Wigham Richardson had also adopted the 'forced' line, with the Neptune yard having 'something like the appearance it kept for over fifty years'. T. McIver, of Swan Hunter's main yard at Wallsend, remained of the view that the arguments for prefabrication and welding could still form the basis of 'an extremely controversial paper'. Thus thirteen years after Hodges's paper had been presented, and with thirteen years of boom conditions behind them, British shipbuilders could still not agree on any form of strategy. *Fairplay* could devote, in 1961, a supplement to Scott's of Greenock entitled 'Shipyard Modernisation' which reported that Scott's had launched its first all-welded ship, a 18,255 dwt tanker, the *Caltex Edinburgh*, in 1956. The trade press could also laud modernisation at Clydebank—'Flow Line Production at Clydebank'—as if a form of revolution were taking place, but Clydebank was merely arriving where most Continental and Japanese yards had been for years.[45] When Professor Roland Smith, then Professor of Marketing at the University of Manchester Institute of Science and Technology, presented a report to the SRNA on marketing, the reaction was predictable. Markets, British shipbuilders believed, were not amenable to long-term analysis, the view being that short-term factors such as freight rates and trade volumes determined the markets. As a consequence of this attitude, collating market information and other related research was considered unnecessary. Rejecting Smith's report in 1969, the Harland and Wolff Chairman, Sir John Mallabar, dismissed market research on the basis that 'if you get an explosion in population, you must get an explosion in world trade. This is all I need to know.'[46] It was a blunt view but probably accurate, at least in terms of how British shipbuilders viewed the world and the market.

Just how Mallabar squared this confidence with the fact that the Belfast shipyard had been consistently unprofitable since 1964 is unknown, but in this respect it was not alone. As the easy years of the seller's market and cost-plus contracts gave way to a buyer's market and fixed-price contracts, the comfortable profits of the 1945–58 period increasingly became losses. The 1958–61 recession brought into sharp relief the increasing price uncompetitiveness of British shipbuilding, and launchings fell from 1.4 mgrt in 1958 to 928,000 grt in 1963. The order book also slid from 5.4 mgrt to 2.1 mgrt between the same dates. Britain's share of world launchings had fallen from over 15 per cent to under 11 per cent, and employment in shipbuilding, ship repairing and marine engineering fell from 297,000 to 222,000. Over two dozen yards were either empty or working on their last orders. Confronted by falling profitability, British shipbuilders blamed surplus world capacity for forcing prices down and generally railed against 'unfair' foreign competition. The stance echoed that of the inter-war years, but the context was that of world economic boom not a depression. This time there was no NSS to buy assets and provide a financial escape route, and the response of many yards was to seek voluntary liquidation. Some of these were heavily ironic in the extreme. On the north-east coast, for example, the closure of the William Gray yard at Hartlepool was especially poignant as William Gray, in combination with Marcus Samuel, the founder of Shell Oil, were largely responsible for the innovation of the oil tanker, with Gray's building eight between 1892 and 1895. Despite this, the tanker did not become a Gray specialism and the yard continued to build a mix of cargo liners, tramps and short sea traders despite the experience of having built standardised Empire ships during the Second World War. Even more ironic was the closure of the Blythswood Shipbuilding Company on the Clyde, which had been specifically laid out to take advantage of the tanker market. Despite considerable success in the post-1945 market, by 1961 the outfitting departments were only being held together by constructing caravans and mobile homes and the yard eventually closed in 1964, citing space restrictions for its inability to win new orders. Another yard to close in 1964, which would have given the lie, to the shipbuilders, at least, that specialism was the wave of the future, was the Clyde dredger specialists Simons and Lobnitz at Renfrew. In general terms, British shipbuilding failed to cope with the trends towards ever larger ships nowhere more so than in the post-Suez crisis closure of the canal in 1956–57.

The significance of the Suez closure, the response of the shipowners, combined with the general recession in shipping between 1958 and 1961, asked British shipbuilders questions which they failed to answer. Penetration of the British market for tankers was barely discernible before 1958, but from that date Britain became a heavy importer of tanker tonnage, becoming a net importer in terms of tonnage launched for the domestic market in 1961 and a net importer in terms of tanker tonnage launched in 1962. The failure in the tanker market, therefore, does explain, in some part, as Table 27 demonstrates, the difficulties faced by the British shipbuilding industry.

Table 27: Tanker launches UK, 1958–1963 (000 grt), and tonnage imported as a percentage of total launched

Year	Total registration	Domestic registration	Overseas	Imported
1958	577	396	181	35.2
1959	529	479	50	72.9
1960	618	534	87	51.8
1961	393	320	74	87.8
1962	403	372	31	106.7
1963	369	233	136	92.4

Source: Calculated from *Lloyd's Register of Shipping.*

As the report of the Committee of Inquiry into Shipping noted in 1970, the closure of the Suez Canal had 'hastened' the speed of development of larger tankers'.[47] The normal measure of the output of tankers used in the market was a measure of comparison with that of the standard T.2 tanker built in the Second World War.[48] This measured the deadweight tonnage of the vessel and its sailing speed. For time series figures, the measure is less satisfactory, as although it does measure tonnage and capacity, it takes no account of the increase in the normal sailing speed of tankers. In terms of T.2 equivalents, the deadweight tonnage of tankers increased by 129 per cent and the carrying capacity by 148 per cent. The economies of scale therefore conditioned the growing size of tankers. Even by 1963, when 100,000 dwt vessels first entered service, by transiting the Suez Canal in ballast and via the Cape when full they could deliver oil more cheaply to Western Europe than could a smaller tanker passing through the Canal in both directions. The cost per ton-mile of carrying oil in a tanker of over 200,000 dwt—37 of which were in service in 1969, with a further 169 on order—was less than one-third that of carrying it in a 20,000 dwt vessel. The larger vessel did not require a larger crew and had a capital cost of about £30 per dwt as compared with £90 per dwt for one a tenth of that size. Although analysis of the size range of tanker fleets does not show the UK register as being much different from that of the world register—79.5 per cent of the British register was in the under 50,000 dwt class, with 20.4 per cent in the over 50,000 dwt class, the equivalent figures for the world register being 79.5 and 20.7 per cent—what is certain is that British orders were increasingly going abroad on the grounds of the ability to build, delivery dates and prices.[49] The trend, therefore, was towards the building of larger ships, but of the 27 shipyards examined by the SIC in 1965, only nine had the capacity to build ships over 40,000 grt and only five could build over 65,000 grt. The capacity of the other eighteen yards ranged from 4,776 grt to 38,996 grt.[50] Thus, the lack of attention by British shipbuilders to the main market trends in tankers and the size of vessels may be interpreted as a major contributory factor in the industry's decline. In the wider context, however, the combination of yard closures and rising unemployment in the run-up to a general election led the Conservative Government to intervene with the credit scheme and probably distilled in the minds of the Labour opposition that it simply could not face the prospect of rising unemployment in its heartland seats. As such, one of the first acts of the new Labour Government in 1964 was to announce an independent inquiry into the shipbuilding industry which, at last, broke the cycle of informal inquiry followed by informal inquiry.

6

Death by Inquiry

Geddes and Jockeying for Position

The late 1950s were the context of a remarkable debate in Britain on the supposed causes of its 'slow growth' relative to its European neighbours and the major industrial powers. Many eyes turned to the remarkable growth record of France where indicative economic planning was hailed as responsible for the turnaround in the French economy. In the UK, a traditionally free market Conservative Government went to great lengths to ape the French methods by establishing a National Economic Development Council, a tripartite organisation comprised of Government, employers' organisations and the trade unions, commonly known as 'Neddy'. This overarching organisation was supported by a range of what were known as 'little Neddies' which were responsible for particular industries and sectors. The Conservative Party began this process but it struck a ringing chord in the Labour Party, long committed to economic planning as the answer to all of Britain's economic ills. As British shipbuilding continued its slide from world leader to international insignificance, Government mounted a full-scale attempt to restore the fortunes of the industry. Whether public money could do what private finance had singularly failed to do remained to be seen.

The newly elected Labour Government which took office in 1964 was essentially defined a year earlier in a speech by the party leader, Harold Wilson, at Scarborough. Anticipating victory, the Prime Minister-elect promised a new Britain 'forged in the white heat of a technological revolution' and identified Labour as the party of modernisation.[1] The record of Conservative Governments since 1951 was castigated as 'thirteen wasted years' and was explicitly blamed for the perceived slow growth of Britain in the 1950s relative to its main industrial competitors. Labour consistently contrasted its modernisation credentials with the Conservative Party's defence of the Establishment and privilege. Wilson had once suggested that much of British industry was 'officered from the pages of Debrett' and that:

> We need a shake up in industry. There's still too much dead wood—too many directors sitting in boardrooms not because they can produce or sell, but because of their family background. To make industry dynamic we need vigorous young executives, scientists and sales experts chosen for their abilities—not their connections.[2]

Given the structure of British shipbuilding it was in all probability an idea which shipyard directors viewed with deep apprehension. The Tories had, however, shot their own fox by the late adoption of some measures of economic planning and intervention in industry as a tacit admission that the policies followed in the 1950s had been wrong. If economic planning was the answer to Britain's economic problems then Labour as the party of economic planning stood poised to reap the reward. The British economy was to be transformed through a 'National Plan'. There would be two new Ministries: the Department of Economic Affairs (DEA), which would administer the plan and counteract the stultifying influence of the Treasury on the 'real' economy; and the Ministry of Technology (Mintech) which would be the prime instrument of industrial reorganisation. Mintech eventually achieved control of Government scientific and industrial research by absorbing the Ministry of Aviation, the Atomic Energy Authority, the National Research Development Corporation, and the Department of Scientific and Industrial Research as well as the operations of the Board of Trade and the Ministry of Fuel and Power. The Ministry also controlled the Industrial Reorganisation Committee (IRC), which was charged with promoting industrial mergers in the belief that the small scale of British firms and industries, relative to their foreign counterparts, was part and parcel of the problem for British industry. Given that the vast majority of shipbuilding yards lay within Labour parliamentary seats it was only natural that Labour should be concerned about shipbuilding. But, to be fair, it did view 'modernisation' as a prescription which would arrest decline and re-establish British shipbuilding as a major force in the world. The industrial strategy was not simply about fostering 'new' industries, but also concerned with the revitalisation of 'old' industries.

Shipbuilding was one of the first industries to be considered a suitable case for treatment. This was spurred by a deputation to Japan in early 1965 headed by the new Minister of Shipping, Roy Mason, which toured Japanese shipyards and met with the Japanese Shipbuilders' Association. The deputation visited eight yards, which were all generally larger than their British counterparts, and in many cases where the new or newly expanding yards were modernising they were doing so on the basis of building docks rather than traditional slipways. The average time taken to build a 70,000 dwt ship in Japan was less than six months, far less than in Britain, and costs were also well below British levels. The report also highlighted the high rates of productivity relative to Britain.[3] The short report produced by Mason proved to be a catalyst and the President of the Board of Trade, Douglas Jay, did what all Governments since 1945 had considered, but averred, and instigated an independent inquiry into the industry. The Committee was established under the chairmanship of Reay Geddes, the Chairman of Dunlop, (thus producing what was commonly known as the Geddes Report). It was charged with attempting:

> to establish what changes are necessary in organisation, in the methods of production, and any other factors affecting costs to make the shipbuilding industry competitive in world markets; to establish what changes in organisation and methods of production would reduce costs of manufacture of large main engines of ships to the lowest level; and to recommend what action should be taken by employers, trade unions and Government, to bring about these changes.[4]

Despite the fact that this Committee was notionally independent, it could hardly ignore the Government's disposition that the scale of British industry was itself a problem, and although there is no direct evidence that the Committee was 'instructed', it does seem to have started from the premise that the scale of the British shipbuilding industry was the major problem inhibiting the development of the industry.

Believing from the outset that grouping was necessary, it is hardly surprising that this was the main conclusion reached by the Committee. Indeed it was central to the findings of the Committee that the industry required reorganisation into larger groups. As the report's conclusion put it: 'the resources necessary for competitive shipbuilding are listed: in general these cannot be provided or supported unless yards are grouped'. Whilst not all yards were geographically suited there was no alternative to grouping as 'neither continued evolution along present lines nor the voluntary co-operation between shipbuilders are likely to lead to the early emergence of the shipbuilding groups which we regard as necessary to the competitiveness of the industry'. One of the Committee's other central recommendations was a programme of Government assistance for reorganisation 'based on the prospects of the industry becoming successful in international competition in a growth market'. Financial assistance, therefore, from the Government through a Shipbuilding Industry Board (SIB) was made conditional upon 'reorganisation'.[5] The industry quickly drew the obvious conclusion that without grouping or reorganisation, or at least a commitment to such, the yards were unlikely to qualify for financial help. As a result of this, as a Vickers internal memo made clear:

> there is a great deal of jockeying for position going on and no one seeing clearly what they are aiming to achieve. The danger in this position is that a number of new groupings may appear which do not make the best use of the capacity and the talent which now exists, and could lead to some of the biggest and most experienced builders losing their predominant position, in favour of smaller groupings.

Indeed, the Geddes Report placed Vickers, with yards on the north-west coast of England at Barrow-in-Furness, and on the north-east coast at Walker on the Tyne, in something of a dilemma. Barrow was one of the yards which was termed geographically isolated by the SIC, which suggested that a grouping with Cammell Laird on the Mersey and Harland and Wolff at Belfast should be considered. The Walker yard, on the other hand, presented an altogether different problem as a configuration around the River Tyne seemed logical.

The Vickers dilemma encapsulated the problems of grouping. In respect of Barrow and any potential merger with Cammell Laird and Harland and Wolff, Vickers was blunt:

> Harland and Wolff continue to sustain heavy losses but presumably due to Northern Ireland Government support they continue in business. Cammell Laird continue to sustain losses in their shipbuilding activities and from our contacts with them we see no reason why, within the next two years, at least, they should be profitable. Barrow can continue to be profitable to an average figure of

> not less than £50,000 per year. Any merger would, therefore, tend to destroy Vickers' profitability in this area as the losses sustained by Harland and Wolff and Cammell Laird are much greater than any profit from Barrow.

Vickers was confident that its privileged position with respect to nuclear submarines and the strength of its order book militated against any grouping which compromised Barrow's position. A 'wait and see' policy was therefore recommended as, in what would prove to be a prescient aside, 'we must remember that size and efficiency do not necessarily group together, and this might yet be well illustrated on the Tyne and/or the Clyde'.[6] The view on the Tyne was rather different, as Vickers realised that there was little likelihood of being able to stand aloof. Three options on the Tyne were identified: remain as at present; await approaches from other companies; or take the initiative in opening negotiations for grouping, with the Barrow tactic of 'wait and see' being rejected. Awaiting approaches, however, would probably result in 'coming out second best in the result', and Vickers therefore believed that it should take the initiative with the options being a Tyne grouping or a Tyne/Wear/Tees group. The favoured option was the latter, combining the Walker yard and the engine works at Palmers Hebburn, the Hawthorn Leslie yard on the Tyne, Austin and Pickersgill on the Wear, and Furness Shipbuilding on the Tees. This, it was argued, gave the best blend of what Geddes had termed B, S and M yards. B yards were defined as those building tankers and bulk carriers, S yards were those which built sophisticated ships such as naval vessels, passenger ships and larger ferries all of which involved a large proportion of fitting out work, whilst M yards built multi-deck mixed cargo ships.[7]

The attraction of such a grouping lay in the output mix which could be produced, the strength of the management teams at Vickers, Hawthorn Leslie and Austin and Pickersgill, and the fact that, with the exception of Hawthorn Leslie, the other yards were owned by large groups which might 'be willing to negotiate on similar terms'. Grouping on the Tyne was explicitly rejected in that 'any group would become an enlarged Swan Hunter', which had been moving 'increasingly into the large tanker market which appears to be the least attractive field to pursue'. In a similar vein, a grouping around Doxford's was rejected in that the Doxford Group itself required heavy expenditure and was 'closely tied to an engineering design and manufacturing company which has had a difficult period in recent years'.[8] As it transpired, the course of grouping on the north-east coast would take a different route from that envisaged by Vickers, but at least the company could be credited with a strategic response to the Geddes Report. Remarkably enough, the history of the post-Geddes intervention in the industry would amply bear out the Vickers contention that size and efficiency did not necessarily go together.

In common with the complications on the north-east coast, the difficulties on the Clyde were labyrinthine. Geographically the river was divided between the Upper Reaches yards running from Clydebank up the river into the heart of Glasgow itself and the Lower Reaches yards centred on Greenock and Port Glasgow. One of the first difficulties resided in the fact that there was a substantial wage differential between the Upper and Lower Reaches yards, which the Lower Reach firms, Scott's and Lithgows, strove vigorously to maintain. Thus a single Clyde grouping, despite the ambitions of the SIC, was never really on the cards. The situation was further complicated by the fact that the Lithgow-owned Fairfield yard on the Upper Clyde at Govan had called in receivers in 1965. This, however, did not follow the pattern of voluntary liquidations which had occurred hitherto. Fairfield's had an order book of over £30 million, had just completed a £5 million modernisation scheme and its failure represented a considerable threat to the Labour Government's modernisation plans. The Government, driven by the head of the DEA, George Brown, decided to mount an ambitious rescue scheme involving the recapitalisation of the yard. In this, the Government, trade unions and private capital all contributed to establish Fairfield's as a national 'proving ground' for the industry utilising new working methods and a new approach to industrial relations. The scheme, though, was widely hated throughout the industry and it soon involved the Government in controversy over its approach to shipbuilding. Here was a Government-sponsored industrial 'proving ground' in shipbuilding running parallel to a Government-initiated inquiry into shipbuilding and there was an obvious contradiction. Despite George Brown's attempt to ensure that the Geddes Committee would take special note of what rapidly became known as the 'Fairfield experiment', the SIC would have been placed in a situation, given its remit, of redundancy were it forced to accept that George Brown and the yard's new Chairman, Iain Stewart, had all the answers to the woes of the shipbuilding industry. Geddes himself had been lobbied by the shipbuilders through the person of Sir Maurice Laing of the Confederation of British Industry and had given Laing the impression

Plate 7: The Clyde in the 1950s. As with Plate 1 the narrow and cramped nature of the river is well-illustrated. In the foreground Stephen's is on the right-hand side of the river with Barclay Curle on the opposite bank and Fairfield's further up on the right-hand side.

that 'he was not supporting the scheme, although he was in no position to oppose it'. Geddes, indeed, took the line that he would 'not favour a solution of Fairfield's problems which was other than temporary' and undertook to represent his views to the President of the Board of Trade, Douglas Jay.[9] Consequently, Geddes wrote to Jay seeking an assurance that 'the statements made about the future of the Fairfield's yard do not limit their [the SIC's] freedom to make recommendations about the future of the industry or the part in which Fairfield's may play in it'. In a second letter, a few day's later, Geddes sought further clarification of the Government's view in order to avoid a situation 'which would make it difficult for our recommendations affecting the industry as a whole to cover Fairfield's as well'. Jay's reply was that the Government had entered no commitment with respect to Fairfield's which would in any way inhibit the recommendations of the SIC, in particular with regard to grouping.[10]

Underlying the 'jockeying for position' was the fact that one of the major recommendations of the Geddes Committee was that the Government should provide financial assistance to the industry. Such assistance, however, was to be conditional on the industry reorganising its structure, altering its outlook and improving its performance. Indeed, support was to be conditional 'on the efforts which are made by both sides within the industry towards full competitiveness'. The recommendation envisaged the establishment of a Shipbuilding Industry Board which would initiate, assist and stimulate action within the industry, administer and control Government financial assistance, and provide Government with informed advice on the prospects of British shipbuilding firms. The SIB was envisaged as an interim measure, which would last approximately five years, until such time as the industry was 'turned around'. Significantly, the Geddes Committee stipulated that shipbuilders should be excluded from the SIB, an explicit acknowledgement that the industry had proved incapable of self-reform.[11] Government financial assistance was to be predicated on the idea of grouping individual shipyards into distinctive large entities, particularly around the major river centres of the Tyne, Wear, Tees and Clyde. Financial assistance was to be provided for consultancy services with a view to establishing groups, loans for the process of grouping itself, which would include support for take-overs, additional working capital, finance for the modernisation of plant and yard layouts, and grants to meet transitional losses and shipbuilding credits.

This strategy was envisaged as being degressive, and would only run to the end of 1970. The conditions for assistance, however, were strict and the industry had to meet a stiff range of criteria and follow a timetable of reforms. Overall, Geddes envisaged financial ceilings on assistance as being £150,000 on consultancy, £32.5 million on grouping and an overall total of £35 million to encompass transitional losses and credits. Thus the SIB would be allowed to allocate over £72 million of public money to the industry until the end of 1970. In return, the SIB would have full access to the shipyards and their accounts, financial assistance was not to interfere with the access to or functioning in the private capital market, investment in non-shipbuilding activities was expressly forbidden, and the SIB had to satisfy itself that expenditure would enhance the competitiveness of individual groups and the industry as a whole. The timetable laid down by Geddes was that the SIB should begin operations by the end of 1966, the consultancy process should be completed by the summer of 1967, loans would be disbursed by the end of 1967, and that until the end of 1970 the Board should consider transitional grants and credits.[12] Given the scale of this potential largesse it was unsurprising that the shipbuilders were 'jockeying for position' with the desire to retain individual profiles in direct conflict with the criteria laid down by the SIC for access to funding.

Praiseworthy as the Labour Government's commitment to industrial modernisation was, like its predecessors, it was consistently to encounter the wider problems of the weakness of both the balance of payments and sterling. Indeed, the new Government's initial economic inheritance was the result of the Macmillan/Maudling growth strategy: a thumping balance of payments deficit. Throughout 1964 and 1965, the Government had to resort to borrowing from both the central banks and the International Monetary Fund, and the use of fiscal and other temporary measures to sustain sterling. Nor was the Government, with a majority of only five in 1964, politically secure until the March 1966 general election, which provided it with a comfortable majority. The impact, however, of having two ministries responsible for different branches of economic policy seemed, to one Cabinet minister at least, to be 'pretty crazy'. The result had been a 'system of having two Ministries permanently at loggerheads, the Treasury and the DEA . . . with sufficient tension . . . between them to cancel out both Departments' policies'.[13] The outcome was that whilst the Government could announce that it had accepted the report and recommendations of the SIC, the Shipbuilding Industry Bill, through which the SIB was to be established, was not enacted until June 1967. Confronted by the possibility that the Geddes timetable could be wrecked before the inception of the SIB, the Government decided to establish it even before the legislation was passed. The Board was chaired by Sir William Swallow, with Anthony Hepper, previously an executive director of Thomas Tilling, and Joe Gormley, the General Secretary of the North Western Area of the National Union of Mineworkers, as the other members. The Board, which reported to the Minister of Technology, was 'charged with the task of promoting the ability of the shipbuilding industry in the United Kingdom to compete in world markets'.[14]

Manoeuvring for position on the rivers now began in earnest. On the Clyde, for example, Fairfield's attempted to open negotiations with Yarrow's and Stephen's although the talks went nowhere. Even earlier, and unknown to Fairfield's, the John Brown Chairman, Lord Aberconway, had convened a meeting between Brown's, Connell's, Stephen's and Yarrow's 'to explore . . . whether the idea of a merger of Clyde shipbuilders should be pursued'. This meeting specifically excluded the other two Upper Reaches yards, Barclay Curle and Fairfield's. Barclay Curle was excluded on the grounds that, as a Swan Hunter subsidiary, it was technically a non-Clyde builder, which did rather seem to beggar geographic logic. Fairfield's was excluded, according to Aberconway, on the grounds that although 'it might be necessary politically, or expedient commercially, to bring them in later', in the meantime 'the stronger and firmer minded the rest of the potential group was by then the better the negotiating position with Fairfield's would be'. The desire to break, or at least emasculate, the 'Fairfield experiment' was, therefore, paramount.[15] Coming from the Chairman of the Clyde yard which probably had the weakest financial position of them all—John Brown's had lost over £3 million on the Swedish American liner *Kungsholm* in 1966—and indebtedness from the shipyard to the parent company was running at over £3.5 million in 1966, this was a bit rich. But as Lord Aberconway had informed the shipyard Managing Director, John Rannie, 'no further moral debt from John Brown to Clydebank remains', and he issued the totally open threat that the outcome of the *Q4* contract, the ship which would become the *QE2*, would be 'decisive perhaps for the continuance of Clydebank as a shipyard'.[16] Although everyone at the meeting agreed that the basis for any merger had to be commercial, there was widespread special pleading that whilst 'current results might well be bad, the future was by no means unrosy', although as Aberconway noted: 'Yarrow particularly . . . seemed happy to "go it alone" confident that his firm would be one of the three chosen Admiralty yards'.[17] At a reconvened meeting the following month, with Aberconway privately viewing 'Clydebank affairs' as 'critical and pertinent' he noted that he found the meetings 'tiresome and frustrating . . . [with] virtually no progress . . . made'. He conceded, however, that he had found it 'difficult to make conversation without unduly emphasising John Brown's position'. With the talks getting precisely nowhere, Aberconway drew the conclusion that talking to Fairfield's might not be the 'dangerous' option that it had hitherto seemed.[18]

The problem for shipbuilders on the Clyde, however, as with those elsewhere in Britain, was that they had no effective control of the situation in that standing aloof from the Geddes recommendations and the views of the SIB effectively entailed a form of financial wilderness. The visit of the SIB Chairman, Sir William Swallow, to Clydeside in September 1966 to establish the parameters for an SIB Upper Clyde Working Party went some way to easing the situation in that according to Rannie 'Swallow [was] not keen on the Fairfield experiment'.[19] This set the tone for another round of discussions with the rest of the Upper Reach builders now reasonably confident that Fairfield's was not going to be allowed to drive the terms of the merger and, if anything, they could probably afford the unexpected luxury of attempting to absorb Fairfield's on their own terms rather than the other way around. In February 1967 the SIB Working Party headed by Tony Hepper arrived on the Clyde and promptly declared that all shipbuilding on the Upper Clyde should be organised into one single group. Although the Working Party's report was not produced until July 1967, the whole affair would take another bizarre twist in that Hepper 'indicated his willingness to take the post of Chief Executive of the Upper Reaches if he were asked unanimously'—an interesting formulation of the old adage which, in this case, saw a gamekeeper turning poacher.[20] When the report of the Upper Clyde Working Party was produced in July 1967, it concluded that the merger should embrace all of the Upper Reach yards. With three yards—Fairfield's, John Brown's West yard and Barclay Curle—as the merchant shipbuilding core, Yarrow's would specialise in naval construction, whilst Stephen's and Connell's would be progressively run down with an eventual conversion to ship repair. John Brown's East yard was to be utilised for the fitting-out purposes of the group. The new entity was to be known as Upper Clyde Shipbuilders (UCS).

The Working Party estimated the capital expenditure required for modernisation as being £1 million, potential productivity improvements across the group of between 33 and 70 per cent, and a return to profitability by 1972–73, compared with average losses of nearly £2.4 million in the three years to 1967. These conclusions were based upon a 'surprisingly bright prospect of the volume of work which the new group could secure' by the management consultants P-E, a view which the Working Party 'found no cause for doubting'. On the market readings the assumed product mix was envisaged as being dry cargo ships, refrigerated ships, container ships, bulk and mixed carriers and ferries and dredgers, with Yarrow's specialising in the naval market. Output, it was argued, would rise from the current level of 210,000 dwt to 350–400,000 dwt per annum, the medium range of which would provide a turnover of £36.3 million. At the upper end of the scale, a mix of merchant and naval turnover could reach £56.9 million. The assumptions regarding improvements was based on a somewhat nebulous view of 'better work organisation, methods and equipment'.[21] Difficulties on the Upper Clyde would merely mirror difficulties elsewhere.

Progress in the north-east of England was also fitful. The SIB initially envisaged a grouping which would embrace the Tyne, Tees and Wear, although this ambition was quickly abandoned as the difficulties of reaching agreements along rivers was awkward enough, never mind between them. Sir John Hunter of Swan Hunter set the tone when he met the SIB in December 1966. He argued that while his own company was viable, 'he did not feel able to justify propositions for mergers with other companies which might become a millstone around his company's neck', and that furthermore 'other shipbuilders could not be expected to sell their physical assets for less than their scrap value' but that 'Swan Hunter did not feel at present that they could buy these assets at scrap value and remain viable'. He discussed the progress of merger negotiations, commenting that Redhead's had been offered 22s 6d a share but had asked for 23s 6d, whilst talks with Vickers had collapsed with Vickers 'on a high horse'. Moreover, though Hawthorn Leslie had good facilities and experience of building specialist ships, the company was making heavy losses on other activities.[22]

Just how difficult the negotiations were, and how fragile individual sensibilities could be, was illustrated later when the Redhead's Chairman, G.H.R. Towers, also met with the SIB. His view on the talks with Swan Hunter was that his company had co-operated in good faith, but had 'got nothing in return . . . Swan's had attempted a straight take-over, and he feared that they might ultimately close Redhead's yard.' He believed that Swan Hunter were waiting for Redhead's next financial results, 'which would not be good', in order to make a reduced offer. In response to this, Sir William Swallow suggested that Redhead's should talk to Hawthorn Leslie with a view to a merger which would place both yards in a stronger position with respect to Swan Hunter.[23] Next through the SIB's door was the Hawthorn Leslie Chairman, Sir Matthew Slattery, who told the Board that negotiations with Swan Hunter and Redhead's were running in parallel, only to be told in turn that any sub-groupings which were formed should facilitate the formation of a single Tyne group.[24] The position was further complicated by the SIB proposal that the Tyne builders should consider the Tees-based Furness Shipbuilding Company in the negotiations, as the yard had recently been modernised, but was widely believed to be in serious financial difficulties. As one builder observed, it was 'already such a fearful problem trying to sort out the four Tyneside shipbuilders, that it would be a mistake to complicate it still further by bringing in the Tees'.[25]

The Tyne shipbuilders' fears over the position of Furness seemed justified when the company's results for 1966 revealed losses of £3 million and projected further losses of £3.5 million for 1967–68.[26] A few days later, Sir John Hunter informed the SIB that the four Tyne shipbuilders had concluded an agreement to merge into a single company to be called Swan Hunter and Tyne Shipbuilders Limited. As to Furness, however, 'all four shipbuilders on the Tyne were unanimously agreed against such an association' and 'Sir John Hunter himself could foresee the closure of the Furness yard'.[27] The Tyne grouping was announced to the public in mid-June 1967 and a few months later scored a major success when it won orders from Esso Petroleum for two 240,000 dwt tankers. The difficulty, however, was that the new group needed almost instant investment to enable the ships to be built. The SIB pledged its full support in terms of capital investment, but also used the opportunity thus presented to bring more pressure on the Tyne group to take over Furness. Commenting on a visit to the Tyne by himself and Sir William Swallow, the SIB Secretary, P.B. Hypher, noted that Cliff Baylis of Mintech had 'more or less pleaded' with Sir John Hunter to absorb Furness. If Swan Hunter would agree, 'every possible assistance would be given by the Ministry and the SIB'. Sir John told Swallow he would be prepared to consider taking over the Furness yard, but only on three conditions:

(i) Swan Hunter would buy the Furness Yard but only on a scrap value basis and with money provided by way of a loan from the SIB. He would not contemplate a merger which involved

Furness having some share or say in the Swan Hunter group.

(ii) There would be no pressure from the Government or the SIB for Swan Hunter to merge with the Wear group, in the foreseeable future at any rate.

(iii) They would expect full SIB assistance by way of whatever guarantees, subsidies etc., they needed.

Sir William Swallow, however, pointed out that the SIB would have difficulties in becoming too heavily involved in what would be essentially negotiations between two private companies, and that public money 'should not be used to "pay off" or buy shipbuilding companies such as Furness'.[28]

When the SIB Director, Barry Barker, visited the Tyne group early in 1968 the whole problem of grouping stood revealed, as Barker reported to the SIB:

> The top management still plays its cards very close to its chest and it is difficult to discern whether the new group is much more than a series of financial deals with the plans for the integration of building facilities, management, labour, etc., still to be worked out, or whether indeed there are any such plans at all. Certainly Sir John Hunter had never visited the Vickers yard until after the group was formed. One therefore had the nagging doubt that an empire has been built without too much thought as to how it is to be fused into a commonwealth.

This, however, could probably stand as an epitaph for the whole post-Geddes approach to shipbuilding modernisation, in that grouping, in and of itself, was assumed, on the basis of little more than an inherent belief in supposed economies of scale, to be enough to turn the industry around. Nor had labour relations improved, which Barker blamed on the approach of the senior management whose 'blinkered commercialism . . . makes every negotiation a deal and every deal a *causus belli*'. This was bad enough, but even worse was to follow when Barker happened to remark in the course of a perfectly innocent conversation that SIB loans were at the Government lending rate, at the time 7⅜%, which came as a 'bombshell'. Sir John Hunter claimed that such a rate of interest rendered SIB efforts 'futile' and that had he known this he would never have embarked on grouping, and Ibison, the Managing Director, went so far as to claim that the yards had been 'conned'. Given that this was the senior management of a group which had been dealing with the SIB for over eighteenth months, it would seem astonishing that none of them knew that SIB loans were being disbursed at the Government lending rate. Hunter declared himself 'completely shattered . . . and the conversation ended with another dissertation on the futility of Government's attempts to assist the industry'. In another change of tack, Sir John now declared that the take-over of Furness would only be considered as part of a 'deal' in return for lower interest rates. Barker's conclusion, however, was that the leverage that the SIB could bring to bear outweighed anything that Swan Hunter could marshal.[29]

The situation at Furness, however, went from bad to worse and by February 1968 the owners, Sears Holdings, had decided upon closure, a course of action that the SIB regretted but saw little alternative to, given attitudes on the Tyne. Neither the SIB nor Mintech, however, had yet conceded the possibility of levering Furness into the Tyne group. Spending two days on the Tyne and at Furness, Cliff Baylis reported back to the SIB that on the Tyne:

> The management is very 'inbred' and seems likely to remain so. All the executive directors come from within the component companies and there is less specialism than would seem desirable. Mr McIver is the key figure: all important decisions are taken by him and he is clearly a man of immense skill, shrewdness and judgement, if somewhat 'old fashioned' in many of his attitudes.[30]

On Furness, Bayliss commented that it 'was the only yard in the UK which reminded him of Japan', although it still had serious problems with overheads some 30 per cent higher than on the Tyne. Whilst Charles Clore, the Sears Chairman, 'would come to terms with Sir John Hunter now: he would not present him with a bargain later'. Bayliss, however, also remarked that he could understand why Swan Hunter did not wish to take Furness over and that the only option would be to 'procure in some way a substantial volume of suitable orders for Furness'.[31] Still, however, the SIB continued to pressurise Swan Hunter, in particular linking the Esso tanker orders to the Furness position. As the SIB commented, the development of a cross-over berth for the construction of the two Esso tankers at Swan Hunter left the yard without space for bulk carriers and OBO (oil/bulk/ore) carriers in the 120,000 to 150,000 dwt range and it was suggested that the Furness yard could fill this gap. With Furness having a book value of £6.4 million at the end of 1967 and being valued by James Barr of Glasgow at £3.48 million on a going concern basis, and £1.325 million on a break-up basis, Swan Hunter made an offer of £1.5 million in cash plus £1 million in shares. The SIB was of the view that this represented the best option short of closure, particularly as the trend in the bulk carrier market was running well ahead of the Geddes expectations. The Board, though, was of the opinion that any funds made available could only be regarded as bridging the gap between what the buyer was willing to pay and the seller prepared to accept. Although it was not totally convinced by the terms of the arrangements, it felt 'that this opportunity to keep available to the industry one of its most modern yards must be taken on a rising market where the best purchaser is willing to act'. The SIB therefore recommended a grant of £0.75 million to Swan Hunter for acquisition of the yard and a minimum grant of £0.25 million, which was to be treated as a minimum, in respect of transitional losses.[32] The Tyne group had been born, although how long its life would be remained to be seen.

As with all other rivers the Wear shipbuilders were unconvinced by the Geddes proposals on grouping, with the Sunderland builders Austin and Pickersgill commenting that with regard to the stated aims of greater efficiency and competitiveness 'we have not been convinced that this would be so'.[33] The Wear Working Party reported in October 1967 and somewhat belied the earlier pessimism by stating that the proposals could produce a viable concern capable of building ships in a competitive market at a reasonable profit. This proposal envisaged a reorganisation around Doxford's with shipbuilding concentrated on three yards, Thompson's, Laing's and Austin and Pickersgill, with Doxford's and Bartram's closing and the new group being known as Sunderland Shipbuilders. The group would concentrate on a product mix of Liberty ship replacements, tramp cargo ships, cargo liners, bulk carriers and tankers in the 10,000 to 100,000 dwt range. Even on this upbeat assessment, however, yet another black hole at the centre of grouping was revealed when the Working Party estimated productivity improvements of over 40 per cent but 'no indication is given of the reasoning, calculations or analysis to show how this figure could be justified'.[34] The Steering Committee report which followed the Wear Working Party proposals argued, however, that both Doxford's and Bartram's should be retained to ensure greater throughput. In all, the Steering Committee believed that the Wear Group would require some £3.345 million to establish.[35] All was not plain sailing, however, and in the summer of 1968 Austin and Pickersgill and Doxford's had a major disagreement over the take-over of Bartram's.[36] As something of a dying fall, in anticipation of his forthcoming retirement, the Doxford Chairman, Sir Henry Wilson Smith, wrote to Barker and attempted to force the issue, arguing that the talks (or lack of them) were stultifying progress and that the only feasible way forward was 'on the basis of a single Shipbuilding Division'.[37] This collapse of what had seemed to be good progress, at least relative to some other areas, brought a full SIB visit down on the Wear in September 1968.

The breakdown of the negotiations on the Wear presented the SIB with severe problems. It revolved around what the SIB regarded as a comparatively simple issue, in that Doxford's already owned Laing's and Thompson's and the only issue to be resolved therefore was in effecting a merger between Doxford's, Austin and Pickersgill and Bartram's. Once again, personal sensibilities intruded on rationality with Doxford's claiming that it was the largest player in the proposed merger, whilst Austin and Pickersgill maintained that it had the most successful and viable product. As the SIB itself admitted:

> The fact remains that by tradition, expertise and topographical limitations, the Wear is the birth-place of smaller bulk and products carriers up to about 25–30,000 tons deadweight, and the simple type of multi-deck cargo ships. Within this range of construction they are world beaters, and as long as a market for these types is available, the Wear has an excellent chance of continuing within the particular market to which it is suited.

The Board cast doubt on the Doxford proposals to increase its potential ship size, and was worried that the single berth Pallion proposal which would entail improvements in the berth, cranage and fabrication facilities would not justify the expenditure. Any expenditure, it was suggested, would be better spent on developing the facilities at both Laing's and Thompson's. The Board noted that both Doxford's and Austin and Pickersgill had returned 'excellent results', Doxford's on a low investment strategy, and Austin and Pickersgill on a high investment strategy. The Board also considered the 'personality' problem and explained, although it did not excuse, the historical reasons whereby 'the Wear shipyards have been and, to some extent, are still in the "family concern and resultant nepotism" department'. As such the history of Doxford's in operating with no integration or rationalisation 'in the true sense' (whatever this meant) was seen as a barrier to Austin and Pickersgill participation. Despite the fact that the ambition remained a single Wear group, the Board had to accept 'that the Wear appears to be much more "off the boil" than other shipbuilding areas'.[38] As it was, the SIB aims were frustrated in that they had to bear two groups on the Wear with Austin and Pickersgill taking over Bartram's in 1968 and Doxford's standing aloof. As with all the others it remained to be seen whether it would work according to the theory, if such is not a grandiose term to apply to the SIB deliberations, and what the outcome would be in practice.

Another of the Geddes recommendations which annoyed the shipbuilders and made grouping more complicated was the conclusion that naval shipbuilding for surface ships should be restricted to three yards.[39] With Vickers and Cammell Laird already enjoying a monopoly of nuclear submarine construction, the clear implication of the SIC report was that the number of yards constructing naval ships would fall from twelve to five. Given the inherent profitability of naval contracts this was a market which no yard and certainly no putative group would be willing to forgo. With Yarrow and Thornycroft having the historic advantage of being specialist naval builders, all the other envisaged groupings fought hard to retain some niche in the naval market. Nor was it made clear when the recommendations of the SIC report were accepted, whether the Government would use its purchasing power as a lever to encourage grouping of shipyards. As Douglas Jay, the President of the Board of Trade, told the House of Commons in August 1966: 'the placing of naval orders will be worked out in consultation with the SIB in the light of the reorganisation of the industry as a whole'.[40] This somewhat Delphic utterance could hardly have reassured the shipbuilders, however, and it meant that the SIB would become a battlefield as new groups strove to retain an interest in the naval market. Nowhere was this more the case than on the Lower Clyde.

Two months before Jay's pronouncement, Michael Sinclair Scott, of Scott's Shipbuilding and Engineering Company Ltd of Greenock, on the Lower Clyde, had been contacted on the implications for grouping by Sir Eric Yarrow and Lord Aberconway of John Brown's. Even at this early stage, however, a Lower Clyde merger between Scott's, which had been a builder of conventional submarines for the Royal Navy, and the mercantile builders Lithgows Ltd, in the adjacent town of Port Glasgow, looked to be a more realistic scenario.[41] Together, their product ranges were complementary and conformed to the SIC criteria for S, M and B yards, although they failed to meet the employment criteria of 8,000–

10,000 workers in a group. Moreover, Scott's was unlikely to be willing to divest itself of its engine building works where it manufactured, under licence, slow speed diesels of Sulzer design. In many respects a Scott–Lithgow merger seemed to be the most natural one on the Clyde. Both firms were geographically isolated and had ready access to deep water with the potential to launch far larger ships than could their competitors on the Upper Reaches. This fact was not lost on the Upper Clyde builders who shared the SIB aspiration of bringing the viable Lower Clyde firms into a single Clyde group.[42] Scott's and Lithgows had also co-operated on labour matters, on the basis that in no circumstances would parity of wages, which were far higher on the Upper Clyde, be conceded. Like many other firms in the industry, family connections remained paramount and although Scott's major shareholder, Swire, kept its stake in the firm as a lock-up investment, the company was chaired by a seventh-generation Scott family member, Michael Sinclair Scott. In Lithgows case, a third generation Chairman, Sir William Lithgow, the son of Sir James, who was also the firm's major shareholder was in sole control.

Before the SIC report was published, Scott's, who also built specialist mercantile vessels, had already acquired two shipyards. The first, the tug, coaster and fishing vessel specialist John Scott and Sons (Bowling) Ltd, was upriver, on the opposite bank of the river, and required an injection of capital to improve its facilities. The second acquisition was the adjacent shipyard and ship repair facilities at Cartsdyke of the Greenock Dockyard Company, owned by British and Commonwealth Shipping, a specialist builder of cargo liners.[43] Lithgows, on the other hand, had lost its Upper Clyde naval yard, Fairfield's, to a combination of receivership, Government rescue and UCS, but had extensively modernised its Kingston yard some years before. The firm's East yard, and the Newark yard of the dredge and specialist vessel builder, Ferguson Brothers, required further modernisation. In the largely derelict former Glen yard of William Hamilton and Company, however, Lithgows had the space to expand its facilities to build the largest type of bulk carriers and tankers, an option which Scott's distinctly lacked.[44] Moreover, both companies were major employers of labour in Greenock and Port Glasgow, both areas of relatively high unemployment.

Three major concerns contributed to make a potential Lower Clyde merger a protracted affair: first, the retention of profitable naval work, in particular the ability to tender for surface warships against SIB and Mintech wishes; second, and inextricably linked to the first, the fear of being pushed into a single Clyde group by the Government and their competitors on the Upper Clyde; finally, and typically, an overriding priority not to put the shareholders' funds at undue risk. All three were very real fears given that the SIB and Mintech favoured a single Clyde group and the obvious lack of confidence which the Ministry of Defence (MoD) had shown in Scott's ability as a builder of sophisticated surface warships.[45] Furthermore, the Admiralty had already taken the decision to go all nuclear in submarine construction, thus limiting Scott's market for conventional Oberon class submarines to the occasional refit or a new order from a Commonwealth country.

Despite the obvious difficulties, however, Scott's and Lithgows announced their intention to merge on 12 September 1967 and had formed a private company, Scott Lithgow Limited, with a nominal capital of £100 as a vehicle to do so. Scott Lithgow Limited was intended to be the holding company, in which the shipbuilding and engineering shares of the two companies would be combined, leaving the parent companies to continue primarily as investment vehicles. It remained fundamental to any real merger, however, that Scott's Cartsburn yard would be designated as the S yard in the Lower Clyde group and, as such, be allowed to tender for surface warship contracts. Consequently, by March 1968, a frustrated Michael Sinclair Scott wrote that he had been trying, with no avail, to receive a complete assurance from Mintech via the SIB that his firm's present role, as a builder of both submarines and surface warships, would continue without hindrance. The SIB had, however, drawn an artificial distinction between submarines and surface warships, conceding the Greenock firm's right to build the former but leaving a question mark over the latter. Scott found this to be 'illogical . . . unreasonable . . . and totally unworkable'. All of this, according to Scott, was directly related to the Government's desire for a single Clyde group.[46] Nevertheless, Scott, equally illogically, wanted a concrete assurance from Government that his firm would be allowed to continue to tender for submarines and surface warships in the future.

The long impasse was finally broken when Scott received a qualified assurance from Anthony Wedgwood Benn, the Minister of Technology, in October 1968, in which he acknowledged that Scott Lithgow could continue to tender for surface warship contracts. Benn later wrote that the Government 'neither possessed nor had sought powers to compel shipbuilding concerns to merge against their will'.[47] Whilst this was true, it was also completely disingenuous as the consequences of not merging were arguably more powerful, with funds at a premium and the prospect of credit applications being delayed. Nevertheless, no Government could bind its successors with an open-ended commitment that any company would be allowed to tender indefinitely. Consequently, the determination of the Lower Clyde to remain outwith a single Clyde group seemed to be totally vindicated, as by June 1969 UCS was virtually bankrupt and was only being sustained by further injections of public money. By this stage, UCS had consumed over £20 million in what was fast becoming a financial 'black hole', whereas Scott Lithgow, in large part due to the protracted nature of the merger negotiations, had received under £20,000 from the SIB as a contribution to consultants' fees.[48] It was plain, even at this stage, that with far more Labour MPs in Glasgow and its environs than the two on the Lower Clyde, the old adage that 'Glasgow is the Clyde and the Clyde is Glasgow' was never more apt.

In all of this, it remained unclear just how Scott Lithgow could hope to compete against a specialist frigate builder such as Yarrow, although some relief was gained in January 1970 when Scott Lithgow won an order from the Government of Chile for two Oberon class conventional submarines. This order prompted Michael Sinclair Scott to state, apparently without a trace of irony, that it 'firmly established Scott Lithgows position among the leading naval builders in the UK'.[49] By the following month, Scott's and Lithgows had finally financially merged their shipbuilding interests through seven private operating companies, previously registered in Scotland for taxation reasons, whose shares, but not physical assets, were held by one holding company, Scott Lithgow Ltd. This coincided with an invitation from Sir Eric Yarrow to Michael Scott to bid for the Yarrow group of companies.[50] These included Yarrow Shipbuilders, in which UCS still held 51 per cent of the equity, Yarrow (Engineers) and the Yarrow Admiralty Research Department (YARD). Given Wedgwood Benn's predilection for a single Clyde group, then an acquisition of Yarrow by Scott Lithgow could have gone a long way to facilitate this. As Scott Lithgow's Managing Director, Ross Belch, informed a senior civil servant, the Yarrow acquisition 'would be a natural first step in this direction'.[51] Thus, after four years of determined opposition to the creation of a single Clyde group, Scott Lithgow's acquisition of the UK's premier frigate builder, with the lure of naval profits, was more than enough for Belch to change tack. It soon became apparent, however, that Vosper Thornycroft was also interested in Yarrow and Sir Eric Yarrow now declared that he desired that any bid for the whole share capital of the Yarrow group should be in cash. With

Scott Lithgow lacking sufficient liquidity to mount a bid of its own, the firm approached H. Clarkson of London, who subsequently agreed to head a consortium to take up shares in Scott Lithgow to enable the cash purchase of the Yarrow group. The question, however, of financial guarantees on contracts given jointly with UCS, and severally by Yarrow, soon came to light. This hitherto unknown factor unravelled the bid, as commercial sense dictated that the future position of Yarrow (Shipbuilders) could threaten the solvency of the entire Yarrow group. As a result of this, therefore, the proposed deal collapsed in May 1970.[52]

At this stage, Scott Lithgow was still negotiating with the SIB on grants and loans to finance a plan to turn its Glen yard into a facility served by a giant overhead gantry crane to construct super tankers of 250,000 dwt with the option of later expansion. Almost four and a half years after Geddes had reported, the SIB gave Scott Lithgow £1.4 million by way of grants and £2.3 million in loans to complete the facility. This was on the proviso, however, that beforehand the parent companies, Scott's and Lithgows, would provide between them £2 million in working capital for Scott Lithgow to be witnessed by an auditor's certificate.[53] Having had its fingers badly burnt by UCS, Harland and Wolff and Cammell Laird, the SIB and Mintech could hardly bear another repeat performance. In this sense, and after the SIB had turned down its request for a loan interest moratorium, Scott Lithgow had paid a heavy price for the protracted nature of its merger.[54] Still, at last, the Scott Lithgow merger was up and running but like all the other mergers it remained to be seen whether grouping would solve the problems of the industry.

If things had gone 'off the boil' on the Wear they were boiling up on the Upper Clyde. Although paying some lip service to the 'Fairfield experiment', the Upper Clyde Working Party report largely ignored it and the announcement that the Deputy Chairmen of the new company would be Sir Eric Yarrow and Michael Scott confirmed the power of the family firms. The Chairmen of the other Clyde firms would be on the Board as Trustee Directors, but the new Chairman, Managing Director, Marketing Director and Personnel Director would all be drawn from outside, a position which, as Iain Stewart noted, would mean that the two Deputy Chairmen 'would virtually control the group'.[55] Dissatisfaction with the financial details of the merger threw Fairfield's and Brown's back together in terms of negotiating a separate Upper Clyde group. The two companies considered attempting to work outwith the SIB, but given that the SIB controlled the finances, a position which could hardly have been lost on either yard given their parlous financial states, this option was rejected. It soon became clear, however, that the SIB was not going to tolerate anything other than one group on the Upper Clyde. By October 1967 the merchant bankers Barings had tabled the financial terms for the merger to the individual yards and the SIB. The complexities of these were revealed as Mr M. Andrews of Hill Samuel, advising Fairfield took the Govan board through the details. Under the proposals, Fairfield's would receive a cash payment of £350,000 and 32.5 per cent of the equity in the new group. Yarrow's, however, would get £1 million, 20 per cent of the equity and retain 49 per cent of their shareholding. In other words, Yarrow's received £1 million to keep 49 per cent of their shares! Connell's received 5 per cent of the equity and a payment of £400,000, whilst John Brown, with no orders, and Stephen's, widely regarded as technically bankrupt, received respectively 30 and 10 per cent of the shareholding both in return for their assets, at a stroke turning lossmakers into shareholders. With Hill Samuel quoted as having been 'unable to obtain a satisfactory explanation of the terms offered to Yarrow', and noting 'the possibility of Brown's having substantial under recovered overheads', the Fairfield Board commented that it was 'not happy with certain aspects of the proposals'. It was, however, in a complete quandary and concluded, reluctantly, that it was:

> not prepared to bear the sole responsibility for the merger breaking down because it believes that it is in the best interests of the Company and of shipbuilding on the Clyde that the merger should proceed. It has therefore decided that, if the Ministry of Technology, the Shipbuilding Industry Board and the Working Party each endorse the acceptance by the other companies . . . the Board of Fairfield's will unanimously recommend their shareholders to accept.[56]

Despite holding their noses somewhat, Fairfield's was in and Upper Clyde Shipbuilders was officially formed on 30 November 1967.

The new company was established with Tony Hepper as Chairman and Iain Stewart, Sir Eric Yarrow and Sir Charles Connell as Deputy Directors, Michael Scott having been dropped when it was realised that a single Clyde grouping was not now achievable. The other three directors were Jim Stephen of Stephen's, Tom Burleigh, the Vice Chairman of John Brown's parent group, and M.O. Hughes of the management consultants Wallace Attwood. By January 1968 James Duff from English Electric had joined as Production Director, John Starks, the Assistant Managing Director of John Brown's, became Technical Director, and another Brown's Director, R.A. Williamson, became Financial Director. As Stewart dolefully observed, despite the fact that no appointments had been made in Marketing and Management Services 'no Fairfield personnel were represented'.[57] Stewart, however, continued to believe that the Fairfield experiment 'would be fairly represented in the new group' in that his right-hand men, Jim Houston and Oliver Blanford, would secure senior positions, but it was to be a misplaced hope. When Houston applied for the post of Director of Management Services, Hepper, Yarrow, Connell, Stephen, Burleigh and Hughes all voted against his appointment. As Stewart told the local MP, John Rankin, 'everything I suggest is voted down. Everything we stood for is either ignored or reversed'. Stewart resigned from UCS on 11 March 1968, less than five weeks after the last Board meeting of Fairfield's (Glasgow), Blanford resigned on 31 March and Houston on 8 April. The UCS merger had killed the 'Fairfield experiment'.[58] Unlike the groupings on the Tyne, the Wear and elsewhere, the Upper Clyde did present a unique case given the Fairfield situation. Perhaps the worst that could be said of Stewart was that he was consistently naive with regard to the politics of the shipbuilding industry. The combined weight of tradition, history and vested interests was extremely heavy. The claim that the deep-rooted problems of the industry could be solved, and by non-shipbuilders, seemed not only an affront to reason but to dignity. Here were people whose fathers, grandfathers and great grandfathers—Yarrow, Stephen, Connell, Lithgow and Scott—had created shipbuilding on the Clyde, being told by someone with no direct shipbuilding experience not only that were they failing in their jobs but also how they should be doing them. The fact that the whole process was being aided and abetted by a 'socialist' Government made the whole situation worse in an industry where free market values were rampant and industrial relations antediluvian.

It was noteworthy that as the merger negotiations proceeded the family firms quickly coalesced leaving John Brown, as a subsidiary of a bigger company, to negotiate with the anomolous Fairfield's. Stewart, it would seem, either did not recognise the tensions or simply believed that once he had proved the point, then everyone else would stand back and ply him with congratulations. Certainly his utterances on the issue were marked by an air of naiveté somewhat rare in the business world. There was a curious form of wishful thinking that things would be all right; this was apparent after the initial shock, after he had explained the situation to the Shipbuilding Conference, after Geddes had reported, and after he was on the Board at UCS. At every turn he was either wrong-footed or deceived. As Oliver Blanford, who had been for two years the Engineering Director at Stephen's, put it:

> I never underestimated the power of families, so even then I made sure my contract allowed me a get-out if they ever managed to take control of Fairfield's. I had it written into my contract by Stewart that I would not be obliged to continue at the yard if anyone was ever brought into a position of executive authority over me. I thought that Stewart had every chance of succeeding with his management experiment in the yard, but only a one in four chance of beating the local vested interests. I knew they would get back.[59]

And get back the families did, for as Jim Stephen was to put it: 'publicly we made no comment on Fairfield's. What we said in private had better remain private.'[60] Upper Clyde Shipbuilders was up and running with one of the main aims of many of its constituents, the emasculation of the 'Fairfield experiment', achieved. Just how difficult the blend of old and new men running UCS would find the operation would not take long to be revealed.

Initially the runes looked good, with the SIB, which had already sunk over £3.5 million into UCS, commenting that 'the new company has since made excellent progress in securing orders, in reorganising its management structure and resources, and its arrangements with labour'. The Comptroller and Auditor General (C&AG), however, took a more pessimistic view, noting the SIB view that if the five yards were left to themselves, with the exception of Yarrow's 'they would sooner or later cease to exist'. Whilst grouping supposedly improved the chances of competing in world markets, the C & AG's view was that 'the question of whether this could be fully achieved could not be answered with certainty'.[61] By September 1968 the capital investment strategy required the full modernisation of Clydebank, Yarrow and Govan, and was estimated at £17.75 million. This, however, was against a forecast loss for the period February to August 1968 of £2 million and a forecast loss on the seven orders taken since the formation of the new company of £1.3 million.[62] In the year to March 1969, UCS consumed a further £3 million in loans and £1.1 million in grants. The conclusion in the SIB report for

1969—that the position with respect to UCS was 'serious'—disguised a turbulent situation. In February 1969 Hepper, who it should be remembered had left the SIB to become Chairman of UCS, appealed to the SIB for emergency financial assistance, seeking £0.5 million simply to enable wages and creditors to be paid. Despite its earlier enthusiasm, the SIB response to this was not encouraging, in that:

> A mere consideration of the results so far suggests that no further financial assistance can be justified, if only because it appears that the losses which are being and would be likely to be sustained in the future could not be recovered from normal profits. But, unless some improvement in operating performance and a consequent reduction of cash requirements for the future were demonstrated within the next 4 months, the SIB would not be in a position to continue assisting the company on its present scale.[63]

Accordingly, the SIB laid down strict conditions with respect to contract prices and costs, and was only willing to pay instalments to a ceiling of £2.5 million until June 1969. Hepper's estimate was that the minimum requirement for 1969 was £6 million, £2.5 million of which was required immediately. To Hepper, it was essential that the SIB gave the 'necessary financial support' to continue a viable shipbuilding industry on the Upper Clyde and the 'absence of this support can only result in liquidation which all must agree would be a National disaster'.[64] In the event, UCS was forced, on the rejection of its corporate plan, to accept the SIB conditions and seek to devise a new strategy.

At a series of meetings in the last week of April 1969, it was decided to close Clydebank (Brown's), Linthouse (Stephen's) and the UCS head office, and to concentrate production at Scotstoun (Yarrow) and Govan (Fairfield's) as Govan required 'the least capital expenditure'.[65] UCS, however, decided to pressurise Mintech on the basis that P-E Management Consultants had reported that either an orderly liquidation or a major retrenchment would 'require financial support not far short of the full amount required to see the company through to 1972/73 but involving considerable loss of employment'.[66] Doubtless this had an impact, with the Minister of Technology, Wedgwood Benn, conceding that 'relations between the SIB and UCS have completely collapsed because SIB are being obstinate rather than tough, and UCS are just declining to pull their fingers out'. Benn, though, was inclined to the view that 'the industrial consequences of the failure of Upper Clyde would be tragic'.[67] With Benn clearly leaning on the SIB, the situation changed once again, with the Government prepared to put in £5 million with the possibility of another £4 million consequent, yet again, on further restructuring.[68] The haemorrhage of working capital, however, continued unabated and the accounts to August 1969 revealed contract losses of £8.39 million, with £3.55 million in inherited losses, one-third of which was incurred by the *QE2*, and £4.84 million on new contracts.[69] With the accession to power of a Conservative administration committed to a 'no lame ducks' industrial strategy, the future of the group appeared bleak.

The election of a new Conservative Government under Edward Heath in 1970 was probably as much of a surprise to him as it was a shock to Harold Wilson. The election victory brought yet another change of policy tack with Heath 'more like a victorious commander moving into a command post he had stormed than the leader of an Opposition party taking over the reins of Government'. As one commentator noted:

> Heath made it clear that for him inefficient industrialists and stupid shareholders were at least as poor combat material as lazy and greedy unionists. They should be treated with even less consideration, allowing below-par enterprises to drop out, to give place to the efficient and combative. And to help them get ahead, the government should stop propping up the incapable and stay out of the hair of the able.[70]

Although Heath is usually characterised as the architect, the policy had been announced by the former Director General of the Confederation of British Industry and new Minister of Technology (after a five-week career as a Tory backbencher), John Davies, at the Conservative Conference in October 1970. There, Davies declared that he would 'not bolster up or bail out companies when I can see no end to the process of propping them up'. He later told the House of Commons that 'the essential need of the country was to gear its policies to the great majority of people who are not lame ducks and who do not need a hand'.[71] They were words, however, that would return to haunt him, but Davies could hardly have known that the ground had already been prepared for him, at least as far as shipbuilding, and in particular UCS, was concerned.

The outgoing Labour Government had kept UCS afloat but it also had to 'bale out' Harland and Wolff and Cammell Laird with the IRC putting in £6 million to the latter company in a move which saw it become a quasi-nationalised entity.[72] Whether the new Conservative Government would continue support on such a scale remained to be seen. Some moves had, however, already been made. Sir Eric Yarrow, who had been reluctant to join UCS in the first place, decided in 1969 to get Yarrow's out of the merger. As part of his strategy he held a meeting with the Conservative shipbuilding spokesman Nicholas Ridley in December 1969. Ridley famously minuted this discussion, noting:

> Yarrows had suffered grievously through the merger, both in profits and labour relations. They are being dragged down by this vast, loss-making and badly run concern. The best long term solution, is having regard to the politics of the situation as well as the economics:
> (a) Detach Yarrow from UCS and allow it to be independent prior to merging on agreed prior terms with Lower Clyde, or Thornycroft.
> (b) For the Government (Labour or Tory) to bail out the rest of UCS—to write off its debts, sell off Government shareholdings, close one or even two of its three yards, appoint a new chairman, and let it stand or fall on its own. This might cost £10m. but would be the end of the nightmare.[73]

Paradoxically, however, it was the Labour Government which agreed the Yarrow de-merger in December 1969, although it would take time to effect. Even more infamous than his reflections on Yarrow's, was Ridley's 'butcher' memorandum in which he concluded:

> I believe that we should do the following on assuming office:
> (a) Give no more public money to UCS
> (b) Let Yarrow leave UCS if they still want to, and facilitate their joining Lower Clyde if they still wish to do so
> (c) This would lead to the bankruptcy of UCS . . . We could put in a Government 'Butcher' to cut up UCS and to sell (cheaply) to Lower Clyde, and others, the assets of UCS, to minimize upheaval and dislocation.
> (d) After liquidation or reconstruction as above, we should sell the Government holding in UCS even for a pittance.
> At this stage we should confine ourselves to saying absolutely

> firmly that there will be no more money from the Tory Government.[74]

Although John Davies would always deny that the policy on shipbuilding had been decided before the election, the fact that the Ridley memorandum would bear a remarkable resemblance to what actually happened is uncanny.

The new Government's tough stance with regard to industry was first undermined in November 1970 when it gave £60 million to the 'blue chip' engineering firm Rolls Royce, which had drastically underestimated the development costs of the RB211 engine to power the Lockheed Tri-star. Two months later, however, the company reported that it could not pay the wages, owed tens of million to creditors and faced penalties running into millions from Lockheed. Here was an early test of the 'no lame ducks' policy. Initially, Heath seemed inclined to let the world-famous firm go, but when it was pointed out to him that airforces and airlines around the world would be grounded, the Royal Navy paralysed, and in all probability Lockheed driven into bankruptcy, he relented and Rolls was effectively nationalised.[75] The conclusion could therefore be drawn that what was good enough for Rolls Royce was good enough for everyone else. Throughout 1970 the tortuous negotiations over UCS continued with the SIB Director on the Board, A.I. Mackenzie, doubting that the company could be rendered viable. What was rendered viable was Yarrow's, which was de-merged, inheriting the £1.25 million covered berth which was one of UCS's few capital investments, and also £4.5 million through the MoD in working capital. As Sir Eric Yarrow would later remark: 'I had the biggest bottle of champagne in my life when I got out of UCS'.[76] For the rest of the group, however, the future was less than rosy. UCS had won one battle in that it had managed to reverse the Government's 1970 policy of refusing to sign shipbuilding guarantees, but this had been achieved on the expectation of profits by September 1970. Not only did these fail to materialise but Mackenzie conducted an analysis of the company's position in May–June 1970 wherein he concluded that the firm was technically trading illegally. There were only two alternatives: a huge injection of public money, or liquidation.[77]

In June 1971, in an eerie echo of the Rolls Royce situation, Hepper asked Davies to guarantee the following week's wages, and told him that liquidation was certain within days. Davies was in a complete bind in that he could hardly avoid the blame if UCS went into liquidation. He bit the bullet, however, and announced to the House of Commons that:

> The Government's attitude is that this company in its present grouping, saddled as it is with debt, dogged by deficits since its inception, having absorbed and lost some £20 million lent and granted to it under arrangements by the former Government is unlikely to achieve a state of stability and prosperity without having recourse to Government aid. Only such a state will ensure the confidence in the future which is needed by workpeople, customers and suppliers alike. The Government have decided therefore that nobody's interest will be served by making the injection of funds into the company as it now stands.[78]

The liquidators would be in, and Davies's only concession was to announce the establishment of an expert group to consider reconstruction. The day after the announcement in Parliament the liquidator, Robert Courtney Smith, announced that the liabilities amounted to £28 million (later amended to £32 million) and that the assets were now no more than £3 million. Smith needed £3 million immediately to pay wages and suppliers for the following six weeks. The expert group, 'the four wise men' as they became known, reported at the end of July 1970 in an astonishing three pages. Davies relayed the conclusions to Parliament that UCS should be reconstituted around Govan and Linthouse, Clydebank would close and there had to be radical reform of both management and labour practices. Employment would fall from 8,500 to 2,500.[79] The report, however, pulled few punches, declaring that UCS, at its inception, had 'a totally mistaken initial structure' and that the new company had inherited 'a massive drain from an already weak working capital by the absorption in UCS of losses from pre-existing contracts'. It concluded 'that any continuation of Upper Clyde Shipbuilders in its present form would be wholly unjustified and, indeed, could cause serious and more widespread damage. It is important that the lessons of this failure are clear and unambiguous.'[80] What was certainly clear and unambiguous was the reaction of the UCS workforce, which now mounted one of the most significant campaigns in the history of British industrial relations.

The unions organised a new tactic, not a strike or a sit-in, but a work-in. The success, or otherwise, of this has been long debated but there is no doubt that the huge profile achieved, reinforced by the dignity and responsible nature of the campaign, gave Government, liquidator and management pause for thought. All soft-pedalled the situation and the work-in was left to wither on the vine, although its ultimate impact was to preserve shipbuilding on the Upper Clyde when the situation, at least potentially, appeared to be that it could (with of course, the exception of Yarrow's) all disappear.[81] By the middle of 1972 a form of reconstruction had been effected. The Boilermakers' leader, Dan McGarvey, had been instrumental in negotiating a deal with the US oil rig builders Marathon to take over Clydebank and he also bullied the other stewards into acceptance of the Govan–Linthouse deal. Attempts at restructuring continued, with Hugh Stenhouse, lately of the Fairfield Board, initially agreeing to take over, but when he was killed in a car crash on 25 November 1971 the Government asked the British Petroleum (BP) Chairman, Lord Strathalmond, to step in. Given that Stenhouse had already negotiated £35 million for the new company, the signs were propitious. Despite this, the Government sought yet another independent study of the viability of UCS from Hill Samuel (which makes one wonder exactly what the role of the SIB had been). The report from Hill Samuel recommended a reconstitution around Govan with the new entity being known as Govan Shipbuilders. It concluded, however, that there was 'no question of the establishment of Govan Shipbuilders . . . being a proposition which could attract commercial support. The decision whether or not to establish it must, therefore, be judged on other considerations.'[82] As Strathalmond commented on the report, 'shipbuilding on the Upper Clyde must be treated at the outset as a social one, or one in the national interest, or both'. Profitability, therefore, did not enter the equation, and as the new Chairman perceptively observed: 'shipbuilding in much of the world has become more of a competition between Governments on the help to be given, in one way or another, to shipbuilders, rather than true competition between these shipbuilders'.[83] It was an observation which would gather force throughout the 1970s and 1980s.

By 1972, therefore, UCS had been broken up but the impact of the work-in had forced, at least, some retreat from the prescriptions of the Ridley memorandum. Rising unemployment and consequent social concerns had forced the Government into what would become, in Tory political folk memory, a 'U-turn'. Yarrow had been returned to the private sector, reasonably secure in the naval market and with a cash injection of £4.5 million. Marathon had acquired Clydebank for £1.5 million but was seeking £12 million in Government assistance to build oil rigs. Govan had been reconstituted and recapitalised by the

Government to the tune of £35 million, and the former Fairfield yard had been rescued for the third time in 37 years. In the short space of four years, UCS had consumed over £70 million in public money, the vast majority of which had been expended in working capital and meeting losses. Grouping, at least as far as UCS was concerned, had proved a costly failure. The comment by Vickers that size by itself was no guide to either efficiency or competitiveness was amply borne out by the UCS affair. This was hardly surprising given that no economic justification had been provided for groupings beyond the conclusion drawn by the SIC:

> that the attainment of a high level of yard efficiency is likely to depend in most cases on the grouping of a number of specialised yards into a single enterprise, with a central management having the resources necessary to ensure a steady work load, adequate planning and design facilities and all the services which such an organisation will require.[84]

The SIC may well have seen the SIB as performing the function of the economic analysis of grouping, or at least having the information available through consultants. Despite this, it appears that the SIB, understanding its remit as being to implement the Geddes proposals, saw grouping as an end in itself. Certainly, the shipbuilders themselves were fixated in that grouping was the necessary precursor to access to Government funds. With all of the Clyde yards in various degrees of financial difficulties (Brown's piling up losses, Fairfield's rescued by the Government and Connell's, Stephen's and Barclay Curle functioning at the bare margins of profitability) and the only profitable yard (Yarrow) clearly fearing for its privileged position with respect to the Admiralty if it stayed outwith a grouping, the whole atmosphere of the situation seemed to change. The equations that grouping equalled money, and grouping equalled efficiency, overrode all other concerns, but neither proved to be routes to viability.

The intervention by Government into the shipbuilding industry may well have been necessary and indeed welcome, at least in some quarters. The problem, however, was that the Labour Government tended to view bigger as better in terms of industrial structures. It was hardly surprising therefore that Geddes recommended grouping and still less that the SIB funded it. In the wider context, however, the SIB increasingly found itself maintaining existence rather than sponsoring modernisation. The initial thrusts of the Department of Economic Affairs and the Ministry of Technology with regard to industrial policy withered on the vine of Treasury opposition. It proved inordinately difficult to formulate a Government ministry which could intervene in a purposive way in industry to any good effect. Intervention, however, would increasingly mean the sustenance of the industry rather than its reform. This would set the tone which would lead to the nationalisation of the industry as a last resort in terms of survival.

7

Things Fall Apart

The 1960s and 1970s were a period of profound change in the maritime industries of the world. As trade boomed, new transport systems, new trade routes and new vessels would emerge to revolutionise the maritime world in as profound a way as steam and steel had in the nineteenth century. The semi-standard break-bulk freighter which had long dominated the world's carrying trades gave way to a new range of specialised ships. There was nothing new in this and it was a widely held view among British shipbuilders that they could build anything. Whether they could build it on time, within budget and to specification, however, remained live issues. More worryingly the new ships for the new trades—the container, the roll on–roll off (Ro-Ro), the lift on–lift off (Lo-Lo), the lighter aboard ship (LASH), the pure car carrier (PCC), liquid natural gas carriers (LNGs) and the oil/bulk/ore (OBO)—were all added to the VLCC and ULCC, but they too seemed to pass British shipbuilders by. Such moves were enhanced when the world economy was hit by the quadrupling of oil prices in 1973, usually referred to as OPEC 1. The impact shattered the 'long boom' and the tanker market collapsed. British output fell precipitously and the Labour Government decided to nationalise the industry as the only way of maintaining it. Whether the State could drive through policies which would help the industry adjust to an entirely new market remained, however, questionable.

As the SIC report noted, 'the present brisk demand for ships has occurred after a recession and there appears to have been a time lag in adjusting to the new situation'.[1] Indeed the market was buoyant, beyond the Geddes estimates, and the combination of the Credit Scheme and the recovering market in all probability served to relax the minds of British shipbuilders. As Table 28 shows, there had been a strong recovery from the mild recession of 1958–61. Yet again, as had consistently happened since 1945, in a period of high demand, British builders

Table 28: UK and world ships delivered, 1962–1971 (000 grt)

Year	UK	World	UK as % of world
1962	1.0	8.2	12.4
1963	1.1	9.0	12.1
1964	0.8	9.7	8.3
1965	1.3	11.8	10.9
1966	1.1	14.1	7.6
1967	1.2	15.2	7.8
1968	1.1	16.8	6.2
1969	0.8	18.7	4.4
1970	1.3	21.0	6.3
1971	1.2	24.0	5.1

Source: Booz-Allen and Hamilton, *British Shipbuilding 1972* (1973).

had settled for a steady state approach to the market. As world demand boomed, British output stagnated in relative terms. The Geddes Report had conjectured three potential outcomes on the basis of action or inaction by the industry with respect to their recommendations. Following full acceptance and implementation, the SIC authors believed that the industry's output could reach 2.25 mgrt, thereby 'justifying the support of the Government'. This growth strategy envisaged the UK industry achieving and holding over 12 per cent of world output. A 'holding on' scenario forecast some changes in attitudes, practices and resource deployment, but essentially a defensive stance with the industry neither renewing nor attracting additional resources. In this scenario, market share was envisaged as 10 per cent with annual output at 1.75 mgrt. The decline scenario posited 'gradual evolution on recent trend' with no gain in competitiveness, no improvement in turnover, falling profitability and cash resources rapidly becoming inadequate. On this basis, market share would fall to 7.5 per cent and the industry would make 'sporadic appeals for government aid to ease the decline'.[2]

In yet another booming market, therefore, British output averaged little more than 1 mgrt per annum. The reasons were not hard to see, a problem of attitude that was comprehensively explained in evidence to the SIC in 1965. Through the Chamber of Shipping of the UK, the Geddes Committee circulated a questionnaire to British shipping companies. Somewhat surprisingly, while the final report was succinct in its summation of the views of British shipping, particularly on issues such as credit, price and delivery dates, it was far less critical than it could have been given some of the material submitted in evidence.[3] The questionnaire drew replies from 27 shipping companies. On the central issues of price and delivery date, fourteen companies expressed the view that prices abroad were lower and delivery dates both quicker and firmer than in the UK. Of the other thirteen firms replying, four did not build abroad as a matter of policy and nine stated that the evidence on prices was, at best, mixed. Three companies—Shaw Saville and Albion, Ben Line and Cunard—were stridently patriotic in their support for British builders, but of the other ten, some reservations were expressed in terms of delivery dates and the quality of the British marine engineering industry. Support for British shipbuilders was strongest from those companies which had long-established bespoke relationships. The Ben Line's relationship with the Clyde yard of Charles Connell, for example, had evolved since 1945 and was based upon a developing vessel design that was subject to incremental improvement. Similarly, the British and Commonwealth subsidiary Clan Line had long favoured the Greenock Dockyard with its contracts. At least one company which did not order abroad, Shaw Saville and Albion, argued that the 'root of all our industrial troubles affecting delivery of home requirements and exports alike' was late delivery by sub-contractors and their failure to honour contracts. Indeed, this was a criticism that was repeated in the replies to the questionnaire—the major failing with British shipbuilding was inadequate control over sub-contractors, leading to late delivery and higher prices.[4]

There was more than a hint, however, that the shipowners were taking their cue on this issue from the shipbuilders, as when the SIC turned to the question of sub-contractors a dramatically different picture emerged. The major problem arising from replies from sub-contractors and ship management companies was 'the failure of some sections of this industry to appreciate that a modern ship is a complex engineering problem', and that there was no longer any room in the industry for yards that were 'a metal box builder' sub-contracting 'all engineering work, both mechanical and electrical, to outside firms'. The shipbuilder of the future had to 'become a co-ordinating organization for all the specialised equipment going into the ship', although the traditional attitude by which the 'shipbuilder built everything and was well able to understand everything fitted on board, dies hard'.[5] While the industry required an increase in the number and quality of engineering staff at all levels, the problem was most acute at top management level. In the view of the Director of PA Management Consultants:

> With some notable exceptions, who could not act independently while still members of the Federation, the calibre of the majority of Shipyard Management was below what one meets in other industries. Backward looking rather than progressive, they are as much responsible as the Unions for the poor state of the industry. Their appreciation of labour relations is poor and conditioned by the past rather than the future. Furthermore, the skill and spirit to bring about the changes which must take place if the industry is not to succumb altogether is lacking.[6]

The problems of sub-contracting, therefore, could not be solved unless the problems of British shipbuilding management were resolved first.

Shipbrokers and consultants may have had one view, but the sub-contractors themselves appealed that they should exert more control over components as the shipyards simply did not possess the engineering skills concomitant with modern shipbuilding. Elliott Marine, for example, had been formed with the express intent of providing complete control systems for main and auxiliary machinery, but had 'encountered considerable resistance from United Kingdom Shipbuilders'. Its view was that production engineering companies should take full responsibility for the engineering, supply, installation, commissioning and servicing of equipment that required 'skills not normally available in UK shipyards'. The small scale of most UK shipyards meant that it was prohibitively expensive for individual yards to attempt to build up or maintain such technical expertise. The Glasgow manufacturer of pumps and cranes J.H. Carruthers and Company Ltd took much the same line, arguing that historically the shipyard had offered all of the technical knowledge that was involved in the design, construction and operation of all types of ships. But, as in all branches of engineering, 'ships . . . are becoming increasingly sophisticated and we believe that the time has been reached now when shipyards should be prepared to pass on to contractors the technical responsibility for large elements of the whole complex, which is a modern marine cargo vessel'. With ships increasingly amalgams of sub-contracted packages, the only way in which a yard could exert more control over the ultimate cost of a ship was through the placement of technical responsibility with firms possessing expert knowledge. The experience of Carruthers, in 'dealing with a number of shipyards outside Britain', was that yards 'which subscribe to this policy . . . today are successful yards'.[7] Despite, therefore, the unanimity of the criticism over sub-contracting, the position seemed far from easily treatable with polar opposite solutions being offered to the SIC.

Even more worrying than the sub-contracting position were the views expressed in evidence to the SIC by a number of shipping companies which had traditionally built in Britain. Alfred Holt and Company, for example, which operated the Blue Funnel Line, and had not ordered outwith the UK prior to 1964, had ordered two specialist cargo liners in Japan in that year. Holt's submitted a lengthy memorandum in support of its decision, claiming that British shipbuilding seemed 'to be in a state of flux and to be deteriorating rapidly', and that although UK prices were generally competitive with most European yards, they were far higher than in Sweden and 20 per cent 'above the Japanese'. The position with regard to delivery dates was termed 'deplorable', with completion times in Japan being at least half of those in the UK. Although it was manifestly more convenient to build in the UK, 'Holt's could conceive of virtually no circumstances in which we would place an order with a British yard'. As if Holt's corporate view were not bad enough, the company's Chairman, Sir Stewart MacTier, argued that the main problem in British shipbuilding was the 'grave deficiency' in the top and middle management in the industry. In his view, there had to be more interest in production planning and production engineering, and this could only be achieved by an influx of graduates and the introduction of top management from other industries.[8]

Another prominent British shipping company, P&O, also began ordering abroad in the early 1960s. P&O outlined thirteen quotations for a 62,000 dwt OBO carrier, four from the UK, five from Europe and four from Japanese yards. British tenders ranged from £3.38 to £3.77 million, European tenders were between £3.32 and £3.62 million, whereas in Japan the range was from £2.90 to £3.24 million. The highest Japanese tender, therefore, was below the lowest British and European tenders. An even poorer position was revealed when prices for an 82,000 dwt carrier were considered, in that the lowest UK quotation was well above Japanese, Danish and Swedish quotations. P&O also expressed concerns over delivery dates, analysing twelve orders that had been placed in six British shipyards in the late 1950s, with only one order completed on time and the other eleven between one and nine months late. Only one vessel had been delivered at the contract price, with the overrun on the other vessels totalling £1.7 million at an average cost of over £160,000 per vessel. Nor had the position altered with the move to fixed-price contracting, in that of four tankers ordered in British yards in November 1962, none had achieved their contracted delivery dates. As P&O noted, British yards themselves had 'little confidence in their ability to maintain forecast dates', whereas they had 'no reason to doubt . . . [that] the anticipated delivery dates offered by the Japanese will be maintained', a fact that was underlined by the 'heavy penalties accepted by the Japanese for late delivery'. At almost all levels—price, delivery dates, technical ability and after sales service—P&O preferred Japanese to British shipbuilders.[9]

As with P&O, BP had normally built in Britain but had started ordering abroad in 1963. BP analysed seventeen quotations for a 19,000 dwt tanker, thirteen from UK yards and four from Swedish yards. The Swedish quotations were between £1.21 million and £1.38 million, whereas the British tenders ranged from £1.49 to £1.91 million. Like the P&O experience, as the size of the vessel rose to the 75,000 dwt range so the price differential between British and Swedish yards widened. Delivery dates were also of concern to BP. In eleven orders placed between April 1963 and January 1964, three in Sweden and eight in the UK, the British orders were between one and ten months late whereas the Swedish orders were delivered on time. British yards took between twelve and sixteen months to construct a 19,000 dwt tanker, whereas Swedish yards took between six and seven months. As BP noted: 'construction periods from placing a contract to delivery are too long and delivery is frequently, if not invariably, late . . . this is deplorable'. As with other companies, BP would have preferred to build in Britain, which it normally did unless no reasonable delivery dates were available. More worrying, however, was its view that 'with the increasing competition in international shipping and the importance of ensuring competitive costs in oil transport, it is doubtful whether tanker operators . . . [would] be able to afford to buy in any except the best (price/quality) market'. In tune with the other commentators, BP had wide-ranging criticisms of the shipbuilding industry, including the 'paucity of higher grade technical staff in the higher echelons of management, but even more particularly at the middle level, where the sheer numbers abroad ensure the attention to detail which is essential to high quality at low cost'.[10] For the British shipbuilding industry, these criticisms were portentous. Given that the industry traditionally relied on the bespoke domestic market, the drift overseas should have been seen as a warning. Added to the fact that the British mercantile marine was growing slowly, British shipbuilders were losing export markets, and now came the ominous indication that British shipowners were increasingly going abroad for their ships, some never to return.

Thus the situation on either side of the recession of 1958–61 was a mirror image: strong world mercantile growth with British output stagnant and a consequent loss of market share. In a situation where world output had tripled between 1962 and 1971, from 8.2 mgrt to 24.4 mgrt, British output rose from 1 mgrt to 1.2 mgrt. Britain's share of world output slid from over 12 per cent to just over 5 per cent. The similarities between the two periods, however, merely disguised major differences. While the Government had worried about the condition of the industry in the 1950s, it had not intervened. The industry was also making comfortable profits and was holding on to the home market. After 1961, intervention had become direct firstly through the Shipbuilding Credit Scheme and then the SIC and SIB. Moreover, the industry had been totally reorganised with a view to international competitiveness. Indeed, grouping had transformed the structure of the industry as Table 29 shows. The geographically isolated yards of Vickers at Barrow, Harland and Wolff, Cammell Laird and Appledore in Devon, were left alone, although the SIB praised their efforts at modernisation which, given that Cammell Laird and Harland and Wolff were quasi-nationalised and Vickers had a near monopoly on nuclear submarine construction, could well have gone down like a lead balloon in Devon. With grouping facilitated public money now flowed like water.

By 1971, the SIB had disbursed almost £43 million to the shipbuilding industry in terms of grants, interest relief grants, loans, consultancy fees and equity, as Table 30 illustrates. The SIB also made recommendations for guarantees by the Government for bank loans in respect of 488 ships with an estimated total contract value of £1,095 million for construction for British owners in British yards.[11]

This was less than half the story, however; indeed, in financial terms it was less than a third of the story, as Table 31 illustrates. Thus, of nearly £160 million provided in assistance to the industry in less than four and a half years, only 34 per cent was provided for capital expenditure. Between them, Cammell Laird, Harland and Wolff (neither of which were in groups), and Govan/UCS consumed over £47 million, with vast sums being expended in writing off losses. Cammell Laird, for example, lost over £12 million on fourteen vessels between 1969 and 1971, while

Table 29: Grouping of UK shipbuilders

Region	Yard pre-SIB	Group post-SIB
East Scotland	Robb Caledon Burntisland	Robb Caledon
Lower Clyde	Scott's Greenock Dockyard Lithgow	Scott Lithgow
Upper Clyde	John Brown Charles Connell Alex. Stephen Fairfield Yarrow*	UCS
Tyne-Tees	Vickers (naval yard) Swan Hunter Hawthorn Leslie Redhead's Furness Smith's Dock	Swan Hunter and Tyne Shipbuilders
Wear	Austin and Pickersgill Bartram	Austin and Pickersgill
	Doxford Laing J.L. Thompson	Doxford
South coast	J.I. Thornycroft	Vosper Thornycroft

Note: * subsequently de-merged.

Source: *Shipbuilding Industry Board*, Report and Accounts, 1971.

Harland and Wolff and UCS similarly consumed vast sums of public money in meeting losses.

In another major difference between the 1950s and 1960s, the easy profits of the former period had evaporated with both inflation and fixed price contracts beginning to bite. As Table 32 demonstrates, only the warshipbuilding sector remained profitable into the 1970s. Whilst the industry itself was quick to blame inflation and sub-contractors for this poor record, it was certainly true that cost increases greatly exceeded the escalation which had been assumed in the fixed-price contracts taken on

Table 30: SIB payments to shipbuilders, 1968–1971 (£000s)

Company	Total
Appledore	1,099
Doxford	13
Govan/UCS	12,792
Harland and Wolff	15,038
Robb Caledon	514
Scott Lithgow	5,246
Vickers	10
Vosper	98
Yarrow	1,573
Others	704
Total	42,925

Source: *Shipbuilding Industry Board*, Annual Reports, 1968–1971.

Table 31: Government assistance to shipbuilding, 1967 to mid-1972 (£000s)

Company	SIB	Other	Total	Percentage devoted to capital expenditure
Appledore	1,099	–	1,099	97
Cammell Laird	–	21,500	21,500	65
Doxford	13	–	13	–
Govan/UCS	12,791	52,300	65,091	19
Harland and Wolff	15,038	36,643	51,681	41
Robb Caledon	514	–	514	78
Scott Lithgow	5,246	–	5,246	72
Swan Hunter	5,838	–	5,838	–
Vickers	10	–	10	–
Vosper	99	–	99	–
Yarrow	1,573	4,500	6,073	15
Other	704	1,900	2,604	20
Total	42,925	116,843	159,768	34

Source: Booz-Allen and Hamilton, *British Shipbuilding 1972* (1973).

Table 32: UK shipbuilding companies net profit (loss), 1966–1971 (£000s before tax)

Company	1967	1968	1969	1970	1971
Appledore	124	127	192	268	217
Austin and Pickersgill	(762)	187	1829	1,829	1,889
Cammell Laird	853	1,354	(7,974)	(628)	(4,275)
Doxford and Sunderland	(28)	1,612	1,238	(2,800)	(1,318)
Govan	*No return*				
Harland and Wolff	(1,156)	(755)	(8,330)	(302)	(182)
Robb Caledon	43	106	(238)	(617)	(504)
Scott Lithgow	312	129	(206)	(1,550)	(262)
Swan Hunter	NA	217	(3,449)	(5,604)	549
Merchant total	(614)	2,977	(16,938)	(9,410)	(3,886)
Vickers	2,533	2,851	2,276	1,474	1,818
Vosper Thornycroft	387	(538)	82	678	643
Yarrow	NA	239	(1,154)	(3,845)	308
Warship total	2,920	2,552	1,204	(1,693)	2,769
Total profit (loss) before tax	2,306	5,529	(15,734)	(11,103)	(1,117)

Source: Booz-Allen and Hamilton, *British Shipbuilding 1972* (1973).

in the late 1960s. Sub-contractors normally bore a major percentage of increased costs, leaving direct labour, miscellaneous supplies and overheads, which normally accounted for approximately 40 per cent of the cost of a ship, under the direct control of the shipbuilders. As one group of consultants found, between 1968 and 1971, from contract date to delivery, inflation on items within the control of the shipbuilders had been as high as 50 per cent, whereas the rate of inflation only produced a 20 per cent increase in total vessel costs.[12]

The differential on losses was accounted for by the perennial problem of the British shipbuilding industry: late deliveries. This was a long standing-problem; as the SIC had commented, 'one of . . . [the] most serious problems' faced by the industry was 'to provide speedy and reliable delivery'.[13] Indeed, the only time in the post-war period when British shipbuilders had been able to meet delivery dates had been in the early 1960s when the order books had been relatively slack. Full, or even fullish, order books had been consistently associated with a poor record of delivery. Despite the repeated claims that this situation could be remedied by increasing capacity or productivity, the industry had proved reluctant to do either. Despite individual modernisation schemes, and a huge amount of Government financial assistance, the position had not improved by the early 1970s, as Table 33 makes clear. This table is based on an analysis of nine merchant shipbuilders, encompassing 287 ships or 88 per cent of all deliveries. Overall, it was found that less than half of the deliveries were either on or ahead of the contract date, with the situation being particularly poor in one area where many had expected substantial growth, that of larger ships. The only companies to be identified as having better than average records were those specialising in smaller ships: Appledore, Austin and Pickersgill and Robb Caledon. These three companies, however, accounted for 40 per cent of ships delivered, but only 12 per cent of total tonnage and 16 per cent of total revenue. Six years on from the SIC report, therefore, 'shipowners report that poor delivery is one of the major reasons for them purchasing in overseas shipyards, where delivery is significantly better'.[14]

Table 33: Summary of the UK shipbuilding delivery record, 1967–1971

Delivery	Percentage of all deliveries
One plus months late	39
Two plus months late	30
Three plus months late	21
Six plus months late	9

Source: Booz-Allen and Hamilton, *British Shipbuilding 1972* (1973).

By 1971, therefore, the British shipbuilding industry was performing even below the Geddes 'worst case scenario', with no gains in competitiveness, no improvement in turnover, falling profitability and cash resources becoming rapidly inadequate. The share of the world market had also slid well under the Geddes 'worst case scenario' of 7.5 per cent of world launchings. Despite grouping, the industry had carried on regardless and had underperformed to an extent that not even Geddes could have imagined. No one, however, could complain that this was due to lack of Government assistance and public money. Both the shipbuilders and the Government now faced a quandary. After years of concern and special pleading the Government had finally acted and public assistance was akin to a leaking tap. But to continue the plumbing analogy, the water was going down the drain. Thus, only six years after the Geddes Report, the Conservative Government was forced into commissioning yet another report into the condition of the British shipbuilding industry through its White Paper on Industrial and Regional Development which was published in 1972. The report was commissioned from Booz-Allen and Hamilton International BV, who were asked to undertake 'an appraisal of the long-term prospects of the shipbuilding industry'.[15] It would, however, be a report which pulled few punches and made Geddes, by comparison, look like a romantic novel.

With the industry performing well below the 'worst case scenario' envisaged by Geddes, and in no manner 'justifying the support of the Government', the report of Booz-Allen and Hamilton, *British Shipbuilding 1972*, made familiar, if thoroughly depressing, reading. The shipowners' views were much the same as in 1965. Late delivery and high prices were now considered endemic, with shipbuilding managements being blamed for deficiencies in marketing effort, labour control and production planning. As the report observed, 'consistent sustained improvement in yard performance is required to alter shipowners' attitudes substantially . . . Certain owners have dismissed the possibility of improvement sufficient to attract them back to UK yards.' Yet again, marketing, or the lack of it, was criticised. Little direct selling was undertaken and such marketing policies as did exist were poorly researched and loosely defined. The main burden of generating orders was still being borne by senior executives, with contract tendering and administration being viewed as support functions. Design remained a problem, with little thought being given to longer-term developments, while the co-ordination of drawing office and production schedules was poorly meshed and resulted in delays in construction. Purchasing and stockholding policies were viewed as essential management services which were, as yet, 'underdeveloped'. Planning remained non-existent, with management neither understanding its role nor recognising its need. As a consequence, there was little co-ordination, with informal approaches to both progress reporting and problem solving remaining the norm. The industry had also failed to develop 'comprehensive personnel and labour relations policies and procedures', while financial planning and control systems were underdeveloped with 'few changes in prospect'. Despite the massive scale of state financial assistance since 1967, there was a continuing reliance on 'traditional methods and old facilities' and 'no evidence of significant improvement in the last five years'.[16] Given that nearly £160 million had been pumped into the industry in less than five years, the Booz-Allen conclusion could barely justify the expenditure.

As world mercantile fleet growth boomed, Britain continued to lose market share. World launches increased from 24.8 mgrt in 1971 to 35.8 mgrt in 1975, whereas UK output crept up from 1.2 to 1.3 mgrt, and Britain's share of world launchings fell from 5 to 3.6 per cent. British launchings in 1971 were almost half the tonnage imported into the UK of over 2.4 mgrt. The views of the shipowners on overseas ordering, so forcefully expressed in 1965, had resulted in the crumbling of the domestic market. As Table 34 makes clear, the scale and speed of this import penetration was astonishing. Despite the claims by the UK shipbuilders that this was largely explicable through 'unfair competition', Booz-Allen argued that 'subsidies, restrictions and credit terms' did not explain the problem, and concluded that 'UK subsidy and export credit arrangements compare favourably with those of other shipbuilding countries, except the USA'.[17] Indeed, under the 1972 Industry Act, the Government undertook to provide 10 per cent of the contract value, and for both home and overseas credit schemes was offering 10 per cent over eight years at a net interest rate of 7 per cent. For a Government supposedly following a robust industrial strategy, this degree of support was astonishing. In two years, Govan consumed £16.6 million of public

Table 34: Ships delivered to the UK registered fleet, 1948–1970 (%)

Year	From UK yards	From foreign yards
1948–50	100	0
1951–55	97	3
1956–60	81	20
1961–65	62	38
1966–70	26	74

Source: *Lloyd's Register of Shipping.*

money, Cammell Laird £6.3 million, whilst Harland and Wolff consumed £20 million from the Westminster Government and £37.6 million from the Government of Northern Ireland. In its four years of office the Conservative Government spent almost £110 million in state support for the shipbuilding industry. Such support, which had initially been targeted on making the industry competitive in the international market, was increasingly being used to avert total collapse. As the Committee of Public Accounts observed in 1972, the vast bulk of money had been used 'simply to meet company losses in order to avoid insolvency'.[18]

With the Government embroiled in industrial strife over its attempt to reform the industrial relations system, and a strike in the coal industry which had led to a three-day working week, the consequent election returned a Labour Government with a manifesto commitment to nationalise the shipbuilding, ship repairing and marine engineering industries. No progress was made, however, between the February and October 1974 general elections and it was not until the end of April 1975 that the Aircraft and Shipbuilding Industries Bill was published. The Bill had a turbulent parliamentary career, being attacked on the basis of a pathological hatred for nationalisation, and for its hybrid nature, by the ideologically opposed Conservative opposition (the same party which had nationalised Rolls Royce) in both Houses of Parliament. This had the support of their friends in the shipbuilding industry, particularly the warshipbuilding firms which mounted a sustained campaign of opposition to it (but still took the compensation). With much of the industry on its knees, however, the Bill, in shredded form, at last reached the Statute Book in March 1977, with vesting day announced for 1 July 1977. Nevertheless, by this stage, Graham Day, who had been earmarked as Chairman Designate in the Organising Committee for British Shipbuilders, and other colleagues had resigned, frustrated by the interminable delays of bringing the legislation to fruition.[19] In what one commentator termed the 'almost endless twilight period prior to full nationalisation', the Government was forced to continue its support for the industry.[20] Sunderland Shipbuilders, for example, which was owned by the Court Line (which also owned the Devon yard of Appledore) went bust. In the summer of 1974 the Secretary of State for Industry, Wedgwood Benn, announced that under the 1972 Industry Act the Government was acquiring the whole of the shipbuilding, repairing and marine engineering businesses of Sunderland and Appledore, at a cost of £16 million, thereby saving 9,000 jobs and securing an order book of over £133 million. Harland and Wolff continued to be an open financial sore. In 1974 the UK Government had taken direct rule over Northern Ireland and by the following year it was announced that the Belfast yard would be nationalised in order to prevent bankruptcy. Between 1966 and 1975, over £81 million had been pumped into the yard and the Government took powers to invest a further £60 million up to 1979.[21] Govan also continued to leach funds, with the £35 million promised over five years by the Conservative Government exceeded by 1975; so the Government put in another £17 million to 1977.

As these various rescues and quasi-nationalisations proceeded, the order book thinned alarmingly as the oil price crisis of 1973, consequent price rises and the collapse of the tanker market all began to bite. Orders in British yards stood at 4.4 mgrt in 1973, but collapsed to only 67,000 grt in 1975. Throughout the industry, yards were being hammered by the unprofitable fixed-price contracts taken on in the early 1970s. This did little more, however, than mirror a general decline of shipbuilding in international terms. The impact of the collapse of the Bretton Woods system of fixed exchange rates, and the 400 per cent increase in the price of oil consequent upon the Arab–Israeli War and resultant sanctions, sent the shipbuilding market into a tailspin. As the glutted order books of the late 1960s and early 1970s worked their way through, world launchings collapsed as Table 35 shows. The first full-blown inter-

Table 35: UK and world shipbuilding launches, 1974–1983 (000 grt)

Year	UK	World	UK% of world
1974	1,281	34,624	3.7
1975	1,304	35,898	3.6
1976	1,341	31,047	4.3
1977	1,119	24,167	4.6
1978	813	15,407	5.2
1979	610	11,788	5.1
1980	244	13,935	1.8
1981	339	17,066	2.0
1982	528	17,290	3.1
1983	527	14,888	3.5

Source: *Lloyd's Register of Shipping.*

national economic crisis since the Second World War had a serious impact on maritime industries throughout the world and it would also mark the beginning of the end of the UK as a volume shipbuilder.

The impact of the oil crisis—the price of a barrel of Middle Eastern oil rose from around $2 in October 1973 to more than $8 by January 1974—combined with supply restrictions and the collapse of the Bretton Woods system, destroyed long-established price, supply and demand schedules and completely undermined the political and economic processes which had been based upon them. In terms of shipping, the market was hit by a range of factors which, in turn, impacted upon shipbuilding. As the price of oil soared, demand collapsed, the tanker market fell apart, freight rates fell and both the spot and charter markets imploded. Moreover, the oil crisis spread throughout the international economy, impacting on general manufacturing, so that the dry bulk market also collapsed and in almost all sectors of shipping the international economy faced massive over-tonnaging and low freight rates. The lay-up rate and mothballing soared and orders were cancelled and ultimately dried up.[22] Doubtless this only hardened the Labour Government's determination to nationalise the industry. Not only was public money flowing into private enterprises, but in addition to the rescue of yards, the cost of construction grants was also soaring. Under Section 11 of the 1972 Industry Act, £1,000 million was available and this was twice raised in 1975 to £1,800 million, but was increasingly being used to cover losses on unprofitable fixed-price contracts. The industry was also hit by the forthright postures struck by other established shipbuilding countries such as Japan, and determined new entrants like South Korea. International shipbuilding confronted its greatest crisis since the slump of the inter-war period and British shipbuilders, in particular, withered under the pressures of a severe and lengthy crisis.[23]

A further complication had been added to the equation when the UK joined the European Economic Community (EEC) on 1 January 1973. European Community (EC) policy towards the shipbuilding industry had been marked by a strong sense of contradiction. Under the Treaty of Rome, no direct subsidisation of military orders was permitted, but the EEC's response to the protracted crisis in merchant shipbuilding was to subsidise in the hope of an improving market while reducing capacity in the expectation that conditions would not improve. As one scholar has noted: 'the two principles are mutually incompatible, but in spite of this they have been pursued simultaneously. It is the emphasis which has changed over time, from the first to the second.'[24] The EC asserted, in

Plate 8: Swan Hunter *c.*1970, a VLCC on the berth. Compare this plate, however with Plate 4.

its various directives on shipbuilding aid, that it wanted a 'healthy and competitive shipbuilding industry consistent with the size of its seaborne trade and respecting its economic, social and strategic importance'. At the same time, however, it wished to recognise the 'imbalance between the production capacity of the industry and demand'. Indeed the EEC advocated the 'adaptation of the structures of the industry to the prevailing market conditions . . . [whereby] production structures should be progressively adjusted to the new market conditions . . . [and this] should be accompanied by measures to facilitate adjustments in employment on a social level'. The *quid pro quo* for subsidy, therefore, was capacity reduction. The directives specifically dealt with 'conversion and closure', particularly the 'social and possibly regional consequences' of such acts, and promised that aid would be available as long as it did not involve the continuation of shipbuilding, conversion or repair.[25] Aid, mainly through the Shipbuilding Intervention Fund (SIF) which was intended to reduce the price gap between Europe and Asia, reached 30 per cent per order by 1984. In practice, however, EEC subsidies tended to operate below the world market price/cost gap and therefore ways around the system were sought, the most common being to use development aid to support ship sales or to utilise other forms of indirect subsidy.[26] The nexus between British Government policy, the framing parameters of European policy, and the international economic crisis, would have a profound effect upon the fortunes of British shipbuilding.

As the Bill to nationalise the shipbuilding industry crawled through Parliament, the scale of the economic crisis facing the maritime industries became more apparent. Remarkably enough, there had been no defaults under the Government guaranteed loan and credit schemes enshrined in the 1967 Shipbuilding Act and the 1972 Industry Act.[27] In 1975, however, the position of the Israeli–American consortium Maritime Fruit Carriers (MFC), which had formed a number of subsidiary companies with UK shipbuilders, and therefore accounted for some 35 per cent of the UK order book at the end of 1975, deteriorated sharply and the Government was forced to intervene. As Swan Hunter, which had formed Swan Maritime with MFC, noted: 'the tanker market to which we were heavily committed with our Swan Maritime partners had completely collapsed' and Swan Hunter observed that the prices being offered by the Japanese were some 50 or 60 per cent lower than the Tynesiders and that they were desperately trying to prevent any cancellation by MFC. The situation, however, was so bad that the Swan Hunter Board 'considered that nationalisation would have some advantages in terms of a long-term assurance of work particularly in the current world shipbuilding market'.[28] Thus, despite the public campaign against nationalisation, there was a developing private belief that it presented a last chance for the industry. Indeed, as the Minister for Trade and Industry, Eric Varley, observed, quoting the *Economist*, a periodical hardly noted for its support of state control of industry: 'public ownership was now the last hope of the industry if it was to survive'.[29] Despite the urgency of the deteriorating situation facing the industry, only in July 1977 did the major firms come under state control.

Under Schedule 2 of the Aircraft and Shipbuilding Industries Act of 1977, twenty-seven companies, comprising sixteen merchant shipbuilding and composite yards, three warshipbuilding yards, five marine engine builders and three training companies, were merged into British Shipbuilders (BS). Subsequently, BS acquired one more merchant shipbuilding firm and six ship repair companies. At vesting day, employment in the group was 87,300, of whom 65,000 were engaged in either merchant or warshipbuilding. Over £75 million in compensation was paid to take the industry into the public sector.[30] BS would have faced a daunting task in whatever circumstances it had been created, given the increasing uncompetitiveness of the industry in a 30-year boom. The new body, with a civil servant, Mike Casey, as Chief Executive, and a former Controller of the Navy, Admiral Sir Anthony Griffin, as Chairman, was confronted by the worst shipbuilding slump since the inter-war years. Moreover, it was initially utterly reliant upon the selfsame yard managements which had brought the industry to the brink of collapse. Although it had been decided that the new corporation would begin its life with a nil commencing capital debt, the scale of what it could expect was revealed in the Swan Hunter accounts for 1976. These showed a loss of over £900,000 after crediting construction grants of almost £3 million, and made loss provisions of £6.4 million for future years.[31] BS, therefore, had little scope in which to deal with the necessary luxuries of evolving a coherent structure and organisation. Beforehand, as the nationalisation bill faced protracted and obdurate opposition, so the international and domestic shipbuilding market went from bad to worse. New orders in 1977 totalled only 429,000 grt, less than half the output for 1977, and the total on order in UK yards at the end of the year was 2.2 mgrt, the lowest end-of-year figure since the early 1960s. In 1978, the world order book slumped to 25.8 mgrt, the lowest level since 1965, and UK new orders totalled 246,000 grt. In this situation, BS's strategy necessarily focused on initial survival rather than any constructive long-term planning effort. Broadly, the attempt was to weather the crisis, the corporation's view being that orders would continue to fall into the early 1980s, but that 'afterwards there would be a rapid climb back towards the levels of demand experienced in the mid-1970s'.[32] In the interim, however, BS took any orders it could get, uneconomic or otherwise. The corporation's first set of accounts revealed losses of over £150 million, loss provisions of some £47 million, and loss provision of another £47 million on orders for 28 Polish ships, an order which filled the books but was wildly uneconomic.[33]

The BS stance was enough to defer the incoming Conservative Government from immediate privatisation, as had been promised in its 1979 manifesto, although continued support was predicated upon rationalisation and a return to viability. With the Government guaranteeing financial support until 1981, the 1979–80 results were disappointing in that the losses amounted to £152 million. The only profit-earners were the designated warship yards, Vickers, Vosper Thornycroft and Yarrow, whilst the largest merchant yards, Sunderland, Govan, Scott Lithgow and Swan Hunter, made losses ranging from £11.7 million to £42.3 million. The new Chairman, Robert Atkinson, decided, in the wake of this, to restructure the corporation. Thus, by the end of 1981, BS had reduced its building sites from 27 to 15; its berths from 66 to 28 and its engine building companies from 5 to 2. Employment fell from over 87,000 in 1977 to 66,000 in 1981, with BS empowered to make special redundancy payments under the Shipbuilding (Redundancy Payments) Act of 1981. Harland and Wolff was explicitly excluded from the nationalisation bill, as it was already a nationalised entity, but outside of BS. It had been granted subsidy facilities of £22 million in the late 1970s, but the trading loss for 1978 was over £21 million with an expected loss of £24 million in 1979. The situation got much worse, and by 1980 Harland and Wolff was reporting that £42.5 million was required just to keep the yard going. In any account, it could hardly be alleged that the Government had not supported the industry. In addition to the money which had poured in during the 1967–77 period, by November 1980 the industry had consumed £316 million in public dividend capital. By March 1981, Harland and Wolff alone had swallowed £206 million, and intervention fund payments to the industry amounted to £105.5 million. Whether Government would continue to support such losses, given the commitment to the privatisation of the industry, was rather more than a moot point given the experience of the Heath Government between 1970 and 1974. But in no industry, however, was the political conflict over industrial policy between Governments more starkly emphasised than in shipbuilding.

One element of Atkinson's new strategy was to diversify the heavy loss-making merchant yards of Scott Lithgow and Cammell Laird into a newly created Offshore Division of BS. There were two reasons for this: first, to take advantage of what was perceived as a strongly growing market (the value of the offshore order book had risen from £39 million in 1978 to £124 million in 1980); and second, to ease the pressure on intervention monies available to the merchant building sector, and to facilitate the strategy of increasing productivity by putting more ships through fewer facilities.[34] Although the strategy had its critics, it also had some history. In 1971, for example, the *Financial Times* had reported a Shell order for four oil rigs at a cost of £32 million, and commented that while the Department of Trade and Industry (DTI) and the major British oil companies had attempted to interest UK shipyards in this work, little enthusiasm had been shown. At both Cammell Laird and, to a much greater extent, Scott Lithgow, a fulsome debate had been conducted at senior management level as to the costs and benefits of diversifying into the North Sea oil and gas platforms and support ship market. The portents for this appeared good, with various reports to Government anticipating a veritable bonanza for British industry due to North Sea oil. The most influential of these was a report from the Scottish Development Department which, in its turn, was based on a report to the DTI by the consultants International Management and Engineering Group (IMEG) in 1972.[35] Although this held out bright prospects, the shipbuilding industry's response to the North Sea gas market in the 1960s had been disappointing. Indeed, of eleven drilling rigs completed in British shipyards between 1961 and 1969, five had been completed at John Brown's (which may explain Marathon's later interest) two at Smith's Dock, and one each at Swan Hunter, Clelands, Furness and Harland and Wolff. Almost all the orders, however, had been taken as book fillers and only Brown's had made any particular virtue of the market.[36]

According to the IMEG report, there would be a need for a minimum of 50 platforms in the UK sector of the North Sea, with 5 to 13 platforms a year required between 1973 and 1983. Assuming a two-year construction cycle, there would be between 10 and 25 platforms or jackets under construction at any given time, with several yards building more than one structure. IMEG also emphasised 'the need for prompt action by the Government', arguing that 'the time for British firms to establish is now or not at all'. The report, however, was far from sanguine about the prospects of the shipbuilding industry in the offshore market. Commenting on information from eight shipbuilding companies, IMEG highlighted the lack of suitable cranage, balanced loading in yard manpower, inappropriate loading for flat panel lines, and inadequate plate rolling capacity, as major barriers to entry for shipbuilding companies. Reviewing the history of rig building in the UK, IMEG stated that it was 'clear that the yards had a very imperfect understanding of what they were undertaking and, with no previous practice as a guide, were widely out in estimating'. The results, therefore, had been either 'a traumatic loss or . . . order books filled with conventional ships . . . [while] only John Brown persevered in the market and other dearly-bought experience has been largely sacrificed'. Consequently 'rig building is clearly regarded . . . as a high risk business and one which does not in any event mix well with shipbuilding in the same establishment'. The report concluded that no British shipyard was likely, of its own volition, to enter the offshore market and that Government, therefore, should hold discussions with shipbuilders to explore the possibilities. It recommended the establishment of one facility, with Government financial aid, to complement Marathon, and that the yard selected for assistance should be indemnified against the possibility of initial losses.[37]

As a result of Atkinson's new strategy, Scott Lithgow won an order from BP for a semi-submersible platform at a cost of £50 million, and an order from Ben Odeco/Britoil for a semi-submersible rig to an Odeco-supervised design in 1981. Cammell Laird gained one order from Dome Petroleum of Canada for a semi-submersible rig. As Atkinson commented: 'this was a major step in consolidating . . . [British Shipbuilders'] involvement in the vast potential offshore market'.[38] The tactic appeared to be working and Atkinson was confident enough to forecast a trading loss for 1982–83 of only £10 million, the loss for 1980–81 having fallen to £41 million, and a break-even position by 1983–84.[39] Against this backdrop, the Government passed the 1983 Shipbuilding Act, which was essentially enabling legislation to facilitate privatisation. Atkinson's optimism as to future prospects was to be rudely dashed, however, when the 1982–83 results, far from showing a further decline in losses, revealed that they had risen to £117 million. Of this total, some £94 million was attributable to only four contracts: three at Scott Lithgow, comprising a tanker and the two oil rigs, and a tanker at Swan Hunter. Accordingly, the Government decided to replace Atkinson with Graham Day, the former Cammell Laird Chairman and previously Chairman Designate of BS, who was given the specific brief of reducing losses and privatising the profitable elements of the industry, the warshipbuilding yards. In terms of the Government's view, this was expressed by the Minister of State at the DTI, Norman Lamont, who declared that 'we cannot afford to have losses on this scale indefinitely . . . the industry is going beyond the Government's funding ability'.[40] The offshore strategy would, however, continue, but in common with a whole range of ironies attaching to the history of the industry in the twentieth century, it would prove the catalyst for privatisation and the ultimate exemplar of bankruptcy.[41]

The brave optimism of 1981 soon gave way to deep gloom as the world shipbuilding market was hit by a second oil price hike, OPEC 2, and Western European governments struggled to adjust their policies to control inflation and tackle debt. As the world maritime industries suffered their second recession in ten years, the impact on British shipbuilding was profound. Table 36 demonstrates the scale of decline. As the market turned down once again so the losses mounted and the adage that competition in shipbuilding was really competition between governments was fully borne out. Nor did the forecasts of a bright future in the North Sea come about as the oil price rises of the 1970s impacted on demand. Indeed, the offshore sector proved to be a disaster, turning in losses which were almost as heavy as those in the mercantile sector, as Table 37 illustrates. Whether the returned Conservative Government

Table 36: UK and world shipbuilding launches, 1984–1992 (000 grt)

Year	UK	World	UK% of world
1984	191	17,732	1.1
1985	145	17,247	0.8
1986	238	14,914	1.6
1987	46	9,770	0.5
1988	91	11,997	0.8
1989	100	13,041	0.8
1990	79	14,894	0.5
1991	151	16,860	0.9
1992	250	20,992	1.2

Source: *Lloyd's Register of Shipping.*

Table 37: British shipbuilders: profit (loss) by specialisation (£m)

	1982–83	1983–84	1984–85
Merchant shipbuilding	(85.5)	(49.2)	(59.3)
Warshipbuilding	54.7	43.8	45.5
Engineering	(2.0)	(12.6)	(8.8)
Ship repairing	(5.3)	(3.7)	(0.8)
Offshore	(76.8)	(100.9)	(3.5)
Other	(2.6)	(38.3)*	(1.7)

Note: * Includes special depreciation provision £37.9m made in 1983–84.

Source: Report by the Comptroller and Auditor General, Department of Trade and Industry, Assistance to Shipbuilders (HMSO, London, 1986).

would support such losses remained to be seen, but there was a developing tendency to view manufacturing industry as being in sectoral decline, blighted by low capital and labour productivity, heavily overmanned with poor industrial relations and overbearing trade unions, and able to survive only through the largesse of the public purse.[42] As Norman Lamont stated in the House of Commons in December 1983, 'the national interest is not to pour good money after bad'.[43]

Given that it was the profitable, but not necessarily more efficient warship yards which were marked down for privatisation, it was richly ironic that the first constituent of BS to be sold off was its heaviest loss-maker, Scott Lithgow. Atkinson had begun the process in 1983 by having talks with the London-based conglomerate Trafalgar House, and despite some expressions of interest from other quarters, Graham Day continued the process with the Government appearing to support the Trafalgar House bid. The offshore contracts, however, were proving to be amongst the most disastrous ever undertaken in British shipyards. In 1983 and 1984 Cammell Laird had run up losses of over £36 million, of which some £27 million was attributable to offshore work, whilst at Scott Lithgow the two rigs being built there had huge loss provisions of over £74 million at the end of 1983.[44] Just why any company with no previous experience of building sophisticated semi-submersible platforms would be interested in purchasing a shipyard in such a parlous state is unclear, especially as it wished to stay in the offshore sector. As Graham Day had observed: 'the financially ill-fated venture into the construction of complete offshore structures . . . [had] impacted upon the Corporation in a number of ways. The financial results speak for themselves.'[45] Perhaps the details of the 'sale' of the yard explain the Trafalgar House interest. Although the headlines claimed that Trafalgar House had bought Scott Lithgow for £12 million, the Government had offered to cover any debts and BS effectively handed £125 million to the conglomerate.[46] It was a remarkable process, but in effect BS had shed its largest single loss-maker and at a stroke the way was paved for the-full scale privatisation of BS. As Graham Day was later to remark: 'I joined British Shipbuilders . . . and moved to get rid of much of the loss making assets as I could, as promptly as I could'.[47]

The result of a second hike in oil prices (OPEC 2) and the election of a monetarist-driven and radically free market-orientated Conservative government, both in 1979, meant that BS hardly had the time to prove its worth before the demands for private operation fell on its head. Despite the fact that privatisation is more associated with Mrs Thatcher's second term in office from 1983, the privatisation of BS was on the agenda from 1979. In the new nexus of industrial policy, essentially defined by membership of the EEC, the government decided to break the industry up. Shipbuilding and the whole culture which surrounded it was regarded as an old industry—in the parlance of the times, a 'smokestack industry'—which UK plc could live without. It was viewed as a bastion of the worst ravages of the post-war settlement: backward management, ante-diluvian trade unions and a poisonous system of industrial relations, all of which sent the wrong message to the wider world. In one sense, however, this was to fight lost battles of yore. In an industry now dominated by computer-aided design and computer-aided manufacturing (CAD/CAM) the image of metal bashing was far from the truth. Furthermore, the flexibility agreements of the late 1970s and early 1980s would have amazed the shop stewards and managements of the 1950s. One question remained, however: could a privatised industry compete in the ferocious market environment of the 1980s and 1990s?

8

Privatised Unto Death

The 1980s witnessed an entirely new approach to economic policy in the UK. Policies were orientated towards the free market, and a form of pseudo-monetarism ruled, reinforced by deregulation and the privatisation of industry. Privatisation, however, was little more than a creeping policy in the 1979–83 period. As the Tories privatised the shipbuilding industry, so companies tried to survive but largely failed. The State-sponsored annihilation of an industry had begun. Whilst the market for ships remained flat, other shipbuilding nations, including those in Europe, performed reasonably well. In the UK, however, the industry did little more than stagnate. The axis between the State and the shipbuilding industry was now broken irretrievably. For the best part of 80 years the State and industry relationships had resolved around trying to revive, sustain or bolster the shipbuilding industry, but now the State simply wanted rid of it. The resources which had been poured in from 1963 had availed little. The projected reforms envisaged by the Geddes Committee had not been realised, and the industry was now left to sink or swim in a new policy environment. In a harsh economic climate, reinforced by a harsh political one, neither of which were clearly readable, shipbuilding was about to be emasculated as a deliberate act of State policy.

Outside the commitment to 'privatise' council housing, the only specific pledges made by the 1979 Conservative Government were to reverse the nationalisation of the aerospace and shipbuilding industries which had been so bitterly contested between 1974 and 1977.[1] The manifestos of 1983 and 1987 were much more robust, citing past successes and future ambitions. Indeed, on industry in general the 1979 manifesto was little more than a promissory note, but one which contained awesome implications in the longer term:

> Government strategies and plans cannot produce revival, nor can

> subsidies ... of course, government can help to ease industrial change in those regions dependent on older declining industries ... there is a strong case for relating government assistance to projects more closely to the number of jobs they create.[2]

By 1983 the trope was to concentrate resources on sunrise industries to the exclusion of sunset industries and combinations of both were marked on the privatisation schedule: sunrise, Rolls Royce and British Airways, for example; sunset, British Steel, British Shipbuilders and British Leyland. It was in this nexus that UK membership of the European Union (EU) would have its greatest impact, although this would take some time to be fully revealed. Although one commentator would later observe that 'British Shipbuilders proved impossible to sell or even give away', this was more or less exactly what did happen.[3] A foretaste of what was to come was revealed by the Chairman, Robert Atkinson, who had outlined his Corporate Plan to 'a very senior official'—actually Sir Peter Carey, the Permanent Secretary at the DTI. Atkinson had been listened to politely, and then told at the end of his peroration: 'Robert, Margaret [the Prime Minister] wants rid of shipbuilding, remember that'. As Atkinson remarked drily, this was 'not very encouraging'.[4] His replacement, however, by Graham Day marked a new approach to shipbuilding. Atkinson had been opposed to privatisation, despite being prepared to offload Scott Lithgow. Day, in contrast, despite his recent tenure of Cammell Laird, was committed to the Government's vision.

As a policy, privatisation was formally announced in July 1984. Initially, however, only the warship yards were marked down for sale, although even this contained a whole series of anomalies. The yards scheduled were the 'traditional' warshipbuilders, Yarrow, Vosper Thornycroft and Vickers, but much more intriguingly the east coast yards of Brooke Marine and Hall Russell, whose only degree of interest in this market was small patrol craft, and, amazingly, Cammell Laird and Swan Hunter. This marked the end of BS's disastrous flirtation with the offshore market but also had the effect of preventing both yards from accessing SIF monies, denied at least to Scott Lithgow in the Offshore Division, and now denied again under the new designation. Exactly how and why the decision was made to designate the seven yards as warshipbuilders is unclear, but with the three protected core warship yards profitable, it may have been done in order to increase competition in the sector with the aim of driving down prices. Whatever, it certainly served

Plate 9: The face of the future. The Doxford ship factory under construction. Compare the new Doxford's with the old in terms of Plate 1.

as a single and costless way of reducing capacity in the merchant sector and putting it 'in a more competitive situation'.[5] As has already been stated, the Treaty of Rome expressly forbade assistance to military shipbuilding, but now a thoroughly hard-headed attitude was also revealed:

> Intervention fund assistance is essentially intended as a temporary aid for merchant shipbuilding and it is provided only for contracts taken on a non-profit basis. Aid of this kind will thus not normally be appropriate for, nor provided to, yards which have been or are to be privatised by British Shipbuilders.[6]

A number of points were clear: once in the warshipbuilding sector there was no way back to the SIF, while private entities coming out of BS would not qualify for assistance.

This also had the curious effect of flying in the face of the policy, first promulgated by Geddes, reiterated by Booz-Allen and confirmed by Lord Carrington in Parliament in 1973, of concentrating naval orders on three yards: Vickers, Yarrow and Vosper Thornycroft.[7] In practice, however, naval contracts had been used as an adjunct to regional policy, treating local unemployment black spots and, in some cases, simply to keep yards open. Thus, although in theory BS had only three warshipbuilding yards, the policy led to confusion, even amongst the senior executives of BS, as to how many warship yards they did, in fact, have. Appearing before the Trade and Industry Committee in 1985, even the Chief and Deputy Chief Executives, Graham Day and Philip Hares, could not agree. On being asked how many warship yards BS had, Hares replied 'five' and Day 'seven'. Questioned directly as to whether composite yards could continue or if warship yards could be redesignated as merchant yards, Hares replied that either would be 'extremely unlikely' as such a move would not fit with the Government's ambition of a 'reduction in the total merchant shipbuilding capacity'. Day offered a further rationalisation of this view, using Swan Hunter as an example:

> Swan Hunter . . . has been a multi-purpose or composite yard, and it is interesting in the sense that it is a competitive (and I am comparing it to the other yards who build warships) warship yard. It is not within British Shipbuilders a competitive merchant yard.[8]

As to the expressed worry that once warshipbuilding was privatised, the rest of BS would, by definition, be unprofitable, this was no more than a fact. Between 1980–81 and 1983–84, the warship division of BS had made a profit of £185 million, as against the merchant sector's losses of £525 million.[9] Day had earlier made use of Swan Hunter to illustrate this point, arguing that 'it had been losing money for years' and that the merchant shipbuilding side of the business was putting the warship-building side 'in jeopardy' to such an extent that it 'could result in that yard ceasing to trade altogether'.[10] Thus, only two types of shipyards were deemed to exist: warship and merchant. The problem for the newly designated warship yards was that while they were not debarred from seeking mercantile orders (although they could not be redesignated as merchant yards) they were denied access to SIF assistance, which effectively meant that any orders were certain to be loss-makers, taken on to fill gaps in the order books.

The policy was also designed, at least if one's logic accepts being stretched to the limit, of facilitating the long-cherished MoD desire for competitive tendering for warships. On this issue, however, as with so many others, the history was lengthy and arguments all too familiar. In 1985 the Committee of Public Accounts expressed concerns over productivity in the yards and the cost of naval vessels, stating that warshipbuilders had 'found it all too easy to make their profits in the past'—no change there as in the past 100 years—and exhorted the MoD to press its 'strong bargaining position'. The MoD refused to believe that genuine competition in warshipbuilding could exist within BS, although to all accounts and purposes the Ministry had never believed that genuine competition existed anywhere, but the Government was prepared to assert that the creation of a competitive base for warship-building was 'a prime aim of privatisation'. The problematical aspect of this policy was that in a sector where it was candidly admitted that 'over capacity . . . currently exists', the policy required, indeed dictated, the creation of a further 'element of over capacity'. As Graham Day saw it:

> at the moment, for large surface warships, the MoD is the key customer. We do not know whether the capacity equals the demand or not . . . that will remain to be seen down the road. At the moment we have responded to the current order position and our anticipation of the future demand by cutting back on a yard by yard basis the manning and the plant and the other infrastructure.[11]

Although he was certainly correct about manpower—employment in the warship yards having fallen from 25,778 in 1977 to 22,235 in 1985—

the statement did not recognise the substantial investments made via public money in now privatised concerns. Vickers (designated Vickers Shipbuilding and Engineering Limited—VSEL), for example, had benefited to the tune of £230 million for the Trident submarine construction hall at Barrow.[12] VSEL also benefitted by acquiring Cammell Laird for £1 in 1985, which gave the Barrow firm renewed surface capability, and a newly modernised construction hall costing £32 million. The cost to BS in loss write-offs was £81.1 million. Despite the protests of the MoD, the favouring of Vickers as the main constructor of nuclear submarines had been little more than a seamless web since the 1960s. The conclusion has to be that the policy was deliberately designed to remove yards from the mercantile sector, while at the same time creating over-capacity in the naval market in an attempt to drive down prices. In the context of a naval strategy based on a 50-ship destroyer/frigate fleet and 27 submarines, plus auxiliaries, this may just have been sustainable, but it certainly implied that any fall in naval orders was bound to create a crisis for the warship sector, particularly given its lack of options.

Still, it was some time before the yards were fully exposed to competition and even the Thatcher Government, for all its much-vaunted free market principles, continued to place orders on a political basis. Swan Hunter, for example, claimed to have 'lost' a Type 22 frigate order to Cammell Laird which, as the Committee of Public Accounts admitted had been placed there to ensure 'the survival of Cammell Laird as a major warshipbuilder', at an extra cost of £7 million.[13] There remained a strong suspicion, however, that in the wake of the 1981 riots in the Toxteth district of Liverpool, the self-appointed 'Minister for Merseyside', Michael Heseltine, had played a disproportionate role in the placing of the order. Swan Hunter were similarly afflicted by 'losing' the order for the first of the new generation of auxiliary oil-replenishment ships (AORs), for which Graham Day had expected it to become the lead yard. Indeed, it had been Swan Hunter which the MoD had commissioned to conduct the feasibility study for these vessels, as even Norman Lamont had argued, it was to the AORs that 'Swan Hunter should look for its future'.[14] As it transpired, the order was placed with the state-owned yard of Harland and Wolff, which had consumed £532.7 million in subsidies between 1975 and 1988. This decision led to the accusation that the order had been diverted to Northern Ireland to keep the yard open and to placate Unionist opposition to the Anglo-Irish Agreement.[15] Amid accusations that Harland and Wolff was cross-subsidising the order (although from which funding source one can only speculate) and that the order made a mockery of the policy of competitive tendering by allowing a Government-owned and hugely subsidised yard to compete with a private-sector yard using only its own resources, it was strange, therefore, that the MoD was admitting that the order should have been placed outside of the designated warshipbuilders. Embarrassed, if such is an appropriate term to apply to Government, the MoD hedged the Harland and Wolff contract with conditional and penalty clauses, and it was conceded that Swan Hunter would be given 'a preferential opportunity to bid for the second class of this ship', but to a Harland and Wolff design![16]

The context of naval-building privatisation was essentially defined by the early defence reviews of the Thatcher Government. Driven by the desire to cut costs across the board, the defence programme could hardly escape. As the Defence Secretary, John Nott, put it in a supplementary Defence White Paper:

> Our current force structure is . . . too large for us to meet the need within any resource allocation which our people can reasonably be asked to afford . . . We cannot go on as we are. The Government has, therefore, taken a fresh and radical look at the defence programme. We have done this in terms of real defence output—the roles our forces undertake and how they should in future be carried out—and not in terms of organisation. It is increasingly essential that we tackle the business of defence in this way, and manage it in terms of total capacity rather than Service shares.[17]

The implication for the Royal Navy was that it was spending too much on ships and not enough on weaponry, and that there should be a consequent reduction in the number of platforms. The subject became one of extremely acrimonious debate which eventually led to the sacking of the Junior Defence Minister, Keith Speed, in May 1981.[18] Again, by implication rather than explication, the White Paper foresaw a radical reorientation of the British fleet away from a surface fleet to an underwater capability with all of the implications that this held for the shipbuilding industry. It was 'saved' (if this is the correct word) by the Falkland Islands 'Emergency', in that the principle of the 'balanced fleet' was resurrected as one of the 'lessons' of the Falkland campaign.

As a result of the Falkland campaign it was accepted that there would be a successor to the Type 22 frigate but that it would have to be significantly cheaper than its predecessor. The new design was initially conceived as being a £70 million ship, but by 1983, with the 'lessons of the Falklands' behind it, the Admiralty had expanded this to £110 million per ship. The three rounds of frigate tendering which took place between 1986 and 1989, exposed the yards to the full force of competition. With Brooke Marine closed in 1986 due to lack of orders and Hall Russell in serious difficulties, the first round of Type 23 orders was split, 2:1 between Yarrow and Swan Hunter, as the MoD remarked 'somewhat uncompetitively'.[19] By the time that the second batch was ordered in 1988, Hall Russell was in receivership, and all three ships went to Yarrow. The Minister for Defence Procurement noted that 'the prices submitted were very keen . . . [and that] less than the average of the last three ordered in 1986'.[20] For the third round, competition was again 'extremely keen' with all three vessels going to Swan Hunter at an 'average unit cost . . . [that was] significantly less in real terms than for the last order which in turn was lower than for the previous ships of the class'. The losing bidders were told to strive 'for better performance to sharpen their competitive edge'.[21] The policy, it seemed, was working, at least as far as the MoD was concerned, but international events were now about to impact on the warshipbuilding sector with devastating results.

Against the backdrop of the fall of the Berlin Wall, the break-up of the Soviet bloc, and the implosion of the USSR, the Government began to revise its defence policy. The immediate fall-out of this and the frigate programme was that Cammell Laird was put up for sale by VSEL. The full implications of having being designated a warship yard were now revealed. As the House of Commons was told:

> A watertight agreement was made, when these yards . . . [the warshipbuilders] were privatised, that they would not be eligible for state aid. The Commission has made it clear that it has no objection to merchant shipbuilding taking place at Cammell Laird, but under no circumstances will it countenance our returning to the Commission and receiving state aid.[22]

Thus, Cammell Laird could return to the merchant shipbuilding sector, but it would not qualify for SIF support, which, in effect, meant that any order which it took could not be economic. Without a military order, therefore, the fate of the Birkenhead yard was effectively sealed. In February 1991 the Government published *Options for Change*, its reaction to events in Eastern Europe, which was widely viewed as the second instalment of John Nott's 1981 review *The Way Forward*. The conclusions for the Navy were, once again, stark, in that the main fleet of 27 nuclear and diesel-powered submarines and 48 destroyers and frigates was to be reduced to 'about' 16 submarines and 40 destroyers and frigates. As the Deputy Under Secretary of State for Policy opined: 'If I was in the shipbuilding industry, I would not regard the prognosis as frightfully good but . . . one of the results of changes in defence requirements is that it does produce some rather unpleasant decisions'.[23] The warship sector lost another 10,000 jobs, as yards attempted to reduce overheads to adjust to the new situation. In the fourth round of bidding for new frigates the whole batch went to Yarrow's 'by a very clear margin'. As the new Minister for Defence Procurement, Jonathan Aitken (who would later end up in gaol for perjury on another matter), rather blandly put it:

> All the changes in the world international scene and in our shipbuilding programme inevitably mean that a reduction is likely in the workload of the industry generated by Ministry of Defence orders. Overall, the potential capacity is in excess of the likely requirements of the United Kingdom warshipbuilding programme.[24]

Not a word was said about the deliberate creation of over-capacity as a means of depressing prices. In 1993, with Cammell Laird closed, VSEL and Vosper Thornycroft secure as the monopoly suppliers of nuclear submarines and small naval vessels, and Yarrow re-established as the predominant yard for frigates, the industry was involved in a straight fight for the few remaining orders, in particular the new landing platform for helicopters (LPH).

In May 1993 it was announced that this order (the vessel would become *HMS Ocean*) would be shared between Kvaerner Govan—a merchant yard—and VSEL; as a consequence, Swan Hunter called in the receivers.[25] Almost immediately, there was a stream of protest that VSEL was in effect 'buying the contract'. This led to investigations by both the National Audit Office (NAO) and Committee of Public Accounts (CPA). Minutes of management–union meetings at VSEL in 1993 revealed that *prima facie* such a charge had substance.

> It was expected that the whole of the MoD current Surface Ship and Submarine Programme could be undertaken by 4/5,000 employees, possibly in one Yard. It was expected that there would

> be a reduction in the number of yards engaged in this work . . . It was important that VSEL/Kvaerner lobbied vigorously for the LPH. Although it was a one-boat programme an elimination of Swan Hunter's would leave VSEL as the only UK Yard capable of building big ships for the MoD.

On frigates, the minutes revealed that:

> This was the most important part of the Surface Ship Programme. The MoD's original plan had been to place a further two contracts for Type 23 frigates embracing 3 vessels per contract . . . in 1994 and 1996. The 1994 batch would not now be ordered and it was expected that only 3 further vessels would be built and would be ordered in late 1995. Should Swan Hunter's remain in business there would be 3 competitors ie VSEL, Swan's and Yarrow's. The CSEU suggested that VSEL had the financial resources to subsidise an LPH bid achieving the objective of Swan Hunter's leaving the field . . . Mr C.N. Davies [the VSEL Chairman] said that it was apparent that without the LPH Swan Hunter could not survive.[26]

Although both the NAO and CPA reports cleared VSEL of 'buying the contract', there is substantial evidence that this was exactly what had happened.

In the first round of bidding for the LPH contract the difference between the two bids was £9 million, with the VSEL bid at £165 million and Swan Hunter's at £174 million. By the second round of bidding, however, with the specifications added, the Swan Hunter bid had risen to £210 million, while VSEL's had fallen to £139 million, leaving a huge gap of £71 million. Despite the admission by MoD officials that no bottom-line comparisons of the two bids had been undertaken by the Ministry, that the 'bids were fundamentally different' and the view that the VSEL bid was being 'supported' from their reserves, the officials could still conclude that the bid was not subsidised. Given that VSEL had £89.5 million in prime profits on the Trident submarine contracts in its coffers, the philosophical difference which MoD officials were able to draw between 'supported' and 'subsidised' may be lost on the ordinary mortal.[27] It availed Swan Hunter little and the yard, at least as a shipbuilder, now entered a protracted death throe. As with all other yards, the Government's favoured solution was to find a buyer, although this ignored two rather fundamental points: first, who would purchase a warship yard from which the Government would not order, and second, who would buy a yard which could not be redesignated in the merchant sector. Some interest, however, was shown in buying the yard by the German company Bremer Vulkan and the French group CMN. But as Fred Henderson of CMN put it, 'the thing that matters . . . is some clear indication from the MoD of its position regarding the future of Swan Hunter as a national shipbuilder . . . and we're not over-encouraged at the moment'.[28] When the Government awarded re-fit contracts to the recently privatised Royal Dockyards, external interest collapsed. The final twists came when, in reply to the campaign group trying to save the yard, Lord Weinstock, the General Electric Company (GEC) Chairman, stated that he was 'afraid we all have to realise that there is too much shipyard capacity for the likely level of future demand'.[29] Two months later, GEC reacted to a take-over bid for VSEL by British Aerospace (BAe) by launching its own bid. There may well have been, as Weinstock had commented, too much shipbuilding capacity, but in the new circumstances Weinstock seemed keen to acquire more.

With two of Britain's major defence companies (one of which, GEC, already owned Yarrow's, as part of GEC-Marconi Naval Systems) in combat over the acquisition of the country's only nuclear submarine builder, the Government had little option but to refer the bids for the consideration of the Monopolies and Mergers Commission. The resultant report admitted that 'since the end of the Cold War there have been substantial reductions in the size of the Royal Navy and in expected orders for warships'. The report conceded that the warshipbuilding sector had been returned, despite the shenanigans of the 1980s and the piety of the MoD's views as to competitive tendering, to the original yards suggested by Geddes as specialist warshipbuilders: VSEL, Yarrow's (GEC-owned) and Vosper Thornycroft. The report also highlighted the fact that the Royal Navy had been substantially reduced in size over the last 40 years and in particular the last 20. Table 38 illustrates the decline. The impact of falling orders and inflated capacity was also noted in that, with prices corrected for inflation, the price of a Type 23 frigate had fallen from the first of class in 1984 to 1992 by some 36 per cent and was 29 per cent less than the first follow-on order in 1986. In a similar vein, the price of a Sandown minehunter platform had fallen by 13 per cent, in real terms, from the first follow-on batch in 1987 to the second in 1994.[30]

The way in which ordering had also contracted was also revealed, as Table 39 shows. In reality, therefore, the naval market, at least in terms

Table 38: Number of ships in the Royal Navy and Royal Fleet Auxiliary, 1975–1995

Type	1975	1985	1990	1995
Submarines	24	24	24	16
Carriers/assault ships	5	4	4	5
Frigates/destroyers/cruisers	59	46	43	36
Mine counter-measure vessels	40	32	40	18
Patrol ships	10	30	32	31
Support ships (including Royal Yacht)	3	2	2	1
Survey ships	13	8	6	6
Tankers	–	13	11	9
Replenishment ships	–	3	3	5
Support and supply	–	2	1	1
Landing ships, logistic	–	5	5	5
Forward repair ships	–	1	1	1

Source: Monopolies and Mergers Commission, 'British Aerospace Public Limited Company and VSEL plc', A Report on the Proposed Merger, Cm 2851 (London, 1995), p. 51.

of builders was back to where it had been before privatisation. As the Monopolies and Mergers Commission conceded, with the elimination of Swan Hunter from naval shipbuilding, no yard 'currently has the operational facilities to build the complete range of ships the MoD may require'.[31] In the event, the Monopolies and Mergers Commission cleared the bids and GEC acquired VSEL. Thus, the major submarine and major frigate yards were under the control of one company with the only possible competitor being Vosper Thornycroft. The fatuity of the situation would be fully revealed in 1999–2000 when GEC decided to restructure and offload GEC-Marconi Marine to BAe. When, in 2000, BAe purchased the Govan yard of Kvaerner, it ended up with a virtual monopoly on the Clyde and a privileged position with regard to warshipbuilding.

The ultimate result of privatising the warship yards was to return the sector to exactly the same size as it had been in the early 1970s, with the then-designated specialist yards surviving. Hall Russell, Brooke Marine, Cammell Laird and Swan Hunter were all closed and employment in the warshipbuilding sector had fallen from 25,778 in 1977 to 10,200 in 1995. Of the four management buy-outs—Brooke Marine, Hall Russell, Swan

Table 39: Royal Navy contracts showing supplier, years of order and delivery, forms of contract and competitors, since 1979

Type	Supplier	Yr of order	Yr of delivery	Quantity	Form of contract	Competitors
SSN (Trafalgar)	VSEL	1981–86	1987–91	4	fixed price	single tender
SSK (Upholder)	VSEL	1983	1990	1	TCIF	single tender
	CL	1986	1991–93	3	fixed price	SL, YSL
SSBN (Vanguard)	VSEL	1981		design	cost plus	single tender
	VSEL	1986	1993	1	FPIF	single tender
	VSEL	1987	1995	1	FPIF	single tender
	VSEL	1990	1997	1	FPIF	single tender
	VSEL	1992	1999	1	FPIF	single tender
Type 42 Destroyer Batch 3	VT	1979	1985	1	FPIF	CL, SH
	CL	1979	1985	1	FPIF	SH, VT
	SH	1979	1985	1	FPIF	CL, VT
Type 22 Frigate Batch 2	YSL	1982	1987	2	fixed price	N/A
	SH	1982	1988	2	fixed price	N/A
Type 22 Frigate Batch 3	YSL	1982	1988	2	fixed price	N/A
	CL	1985	1989	1	fixed price	directed
	SH	1985	1989	1	fixed price	CL, VT
Type 23 Frigate	YSL	1982		design	cost plus	single tender
	YSL	1984	1990	1	TCIF	single tender
	SH	1986	1991–92	1	fixed price	single tender
	YSL	1986	1991	2	fixed price	SH, CL, VT
	YSL	1988	1993–94	3	fixed price	SH, CL, VT
	SH	1989	1994	3	fixed price	YSL, CL
	YSL	1992	1995–97	3	fixed price	SH, VSEL

Table 39: continued

Type	Supplier	Yr of order	Yr of delivery	Quantity	Form of contract	Competitors
Minesweeper/patrol River class	R	1982	1985–96	5	N/A	N/A
Mine Countermeasures Hunt class	VT	1980	1985	4	FPIF	single tender
	VT	1982	1986	2	fixed price	YSL
	VT	1985	1989	2	fixed price	YSL
Minehunter/Sandown class	VT	1985	1989	1	FPIF	single tender
	VT	1987	1991–93	4	fixed price	YSL
	VT	1994	1998–2001	7	Firm/fix	YSL
LSL	SH	1984	1987	1	N/A	N/A
Aviation training ship (conversion)	H & W	1984	1988	1	N/A	CL
LPH	VSEL/KG	1993	1997	1	fixed price	SH
AOR	H & W	1986	1993	1	fixed price	SH, CL
	SH	1987	1993	1	fixed price	Directed

Key:

Shipyards:

CL = Cammell Laird
H & W = Harland and Wolff
KG = Kvaerner Govan
R = Richards (Great Yarmouth and Lowestoft)
SH = Swan Hunter
SL = Scott Lithgow
VSEL = Vickers Shipbuilding and Engineering Limited
VT = Vosper Thornycroft
YSL = Yarrow Shipbuilding Limited

Contract:

FPIF = Fixed price incentive fee
TCIF= Target cost incentive fee

Competitors:

Directed = Contract awarded by ministerial direction

Source: Monopolies and Mergers Commission, 'British Aerospace Public Limited Company and VSEL plc', A Report on the Proposed Merger, Cm 2851 (London, 1995), p. 69.

Hunter and Vosper Thornycroft—only the latter was successful. GEC, which had purchased Yarrow in 1986, also took over VSEL in 1995. The whole process was hailed as a success by the Government, which claimed a gain for the public purse of £120 million. The same purse, however, had paid out £235 million in investment capital between 1984 and 1986 alone, bore the accumulated liabilities of £81.1 million for Cammell Laird, accepted costs of £1.6 million for Hall Russell, and continued to carry contingent liabilities for future redundancies elsewhere. In truth, there was never much prospect that the inclusion of the small yards such as Brooke Marine and Hall Russell, or the redesignation of Swan Hunter and Cammell Laird as warshipbuilders, would fulfil the ambitions of the MoD with regard to privatisation. Certainly competition for orders was fierce, particularly against the background of cuts in naval procurement, and prices were driven down, but new entrants were always unlikely to be able to compete with the long-established lead yards. The cynical suggestion—although it is a conclusion which it is difficult to escape—would be that if the policy had begun as an easy way of reducing merchant shipbuilding capacity, it had ended as a way of eliminating capacity altogether. Once designated for warshipbuilding, and incapable of being redesignated for the purposes of merchant shipbuilding, yards could not access the SIF and therefore could not hope to compete for mercantile orders. With no prospects in the merchant sector and no naval orders forthcoming, there was little alternative for many but closure.

Given the 1979 manifesto commitment to full-scale privatisation of the shipbuilding industry, very few believed that this would be possible in the chronically loss-making mercantile sector. The international market for merchant shipbuilding remained depressed throughout the 1980s. Having recovered somewhat from the nadir of 11.8 mgrt in 1979, the world market expanded slowly to 1982, dipped in 1983, recovered somewhat to 1985, before collapsing to 9.8 mgrt in 1987. In this depressed market, Britain fared poorly. In 1979 the UK launched 610,000 grt or 5.1 per cent of world output; by 1984, however, this had fallen to 191,000 grt or just over 1 per cent of the world total. In the slump of the mid-1980s, UK output slid to only 46,000 grt or 0.5 per cent of world launches. In this fiercely competitive market, the only Western European shipbuilding nations to expand their output were West Germany, Denmark and Finland, while Sweden (so often advanced as the exemplar to British shipbuilding) decided to come out of shipbuilding altogether. By way of contrast, Japan, which reduced capacity in the 1980s by almost one-third, launched 4.3 mgrt in 1979 and 7.1 mgrt in 1984. Even more astonishing was the progress of South Korea, which had launched only 15,598 grt in 1972, 478,718 grt in 1979 and 2.5 mgrt in 1984. The general view was that either BS would be reduced to a small core and privatised when market conditions improved, or as one MP put it: 'when naval shipbuilding . . . [has been] privatised, the rapid decline of British merchant shipbuilding will set in . . . it will be a terminal illness'.[32] It was to be an accurate prediction. Announcing the 1984 BS corporate plan to Parliament, Norman Lamont stated that BS's contract target had been reduced by 50 per cent, and that the new product strategy involved the avoidance of direct competition with East Asia and concentrating on the value-added areas of the market. Appearing before the Trade and Industry Committee, Graham Day stated that the South Koreans were prepared to buy business below the cost of materials, manufacture and overheads, with no regard to profit. He then described just how brutal the process was:

> We know that in the day to day market, if for example you or I were an independent shipowner, and we had a communication from, say, a Japanese shipbuilder that said the price was 10, all you would have to do is produce that to a South Korean yard and they will bid 9 without looking at the specification or anything else.

There was no point, therefore, in even attempting direct competition with the Koreans. Day also charged that in Europe 'we are playing cricket and someone else is playing rugby league and they are getting their retaliation in first', a reference to the use of subsidy and aid packages to circumvent EC directives. BS was, therefore, in 'no position at all'.[33]

There was also, however, a hidden agenda behind Day's appointment which only became public after Day had left the Corporation. In questions about whether Day's successor, Philip Hares, and the DTI agreed on goals, the Permanent Secretary, Sir Brian Hayes, was asked that 'the objectives . . . be stated openly and in public'. Hayes's reply could have been lifted from a script of *Yes Minister*, minus the humour:

> We will certainly consider whether it would be possible to publish an edited version without indicating that it had been edited. If one were to suggest that there were additional secret objectives that could make Mr Hares task of managing British Shipbuilders rather more difficult than it would otherwise be, I ought perhaps to say

> that Mr Day did in fact reveal, with the Department's consent, his own personal objectives to the other members of British Shipbuilders' Board. There was no question of his acting in a covert manner in relation to them, but they were not made public and the same would apply to the objectives agreed with Mr Hares.[34]

The secret objectives were eventually wrung from the DTI by the Comptroller and Auditor General and published as an appendix to a CPA report. Day's aims were to meet financial targets set by the Secretary of State; to pursue a robust programme of cost reductions which would reduce the annual trading loss and eventually achieve a break-even position. Offshore losses were to be eliminated, engine-building capacity reduced and other facilities closed or sold.[35] Day was as good as his private word. Scott Lithgow was sold, Cammell Laird was redesignated as a warship yard, and three yards (the Leith yard of Robb Caledon and Clelands, two yards at Wallsend and Goole) were closed. Financial support to the industry was also reduced. This was sparked by the fact that SIF subsidies of £40.4 million had been provided for sixteen contracts in 1983–84, on the basis that no losses would be incurred. In fact, losses amounted to over £89 million, more than twice the SIF assistance. Accordingly, the Government instituted a new Contract Support Limit which provided that if contract losses were sustained, BS's ability to take on new orders with SIF assistance was automatically limited.[36] In May 1986 the Government announced the closure of Smith's Dock on the Tees, Ailsa Ferguson at Troon and the Wallsend engine works of Clark Kincaid, with the loss of 3,500 jobs.[37] BS was reduced to North East Shipbuilders Ltd (NESL), a combination of Austin and Pickersgill and Sunderland Shipbuilders on the Wear, Govan and Ferguson on the Clyde, Appledore in Devon and Clark Kincaid's engine works at Greenock.

Day had been frank enough appearing before the Trade and Industry Committee in maintaining that BS existed 'as virtually all other shipbuilders exist, with a measure of government support'—not, in all probability, something which the Government wished to hear. When asked whether the only alternative to Government support was yard closures, Day was direct:

> If there was no government support there would not be merchant shipbuilding in Britain . . . if government said to us, 'You must only carry on with those companies which were profitable and stand

> aside on privatisation', we would have no choice but to close all merchant shipbuilders because they are all unprofitable. This has been the case since 1977.[38]

It could, however, have been a statement of policy rather than an answer to a question as standing aside on privatisation was pretty much the tactic. In 1986 Philip Hares, Day's successor, was equally blunt, declaring that if BS 'received only as much this year as we received in the last . . . the industry could fold up'. Caught between a declining Contract Support Limit, and restricted in terms of access to the SIF, it was hardly surprising that the industry was failing to attract orders in an already extreme market. Even more blunt than Hares was the new Minister of State at the DTI, Peter Morrison, who told the Trade and Industry Committee that 'you cannot keep shipyards open doing nothing'.[39] Contemporaneously, however, the international market was further dislocated with the collapse of Sanko Steamships, which flooded the market with 125 new vessels, adding depression to depression.[40] Having made a trading loss of £125 million in 1985–86, BS lost a further £50 million with the collapse of International Transport Management Ltd, for which Sunderland was building a sophisticated craneship.[41] Against this, NESL secured an order from Denmark which seemed to secure the future of the yard but in reality proved to be the agent of its demise.

Exactly when the Government decided to privatise the remainder of the mercantile sector is unclear, but it was very likely established policy during the Day tenure. In any event, yet another new Minister of State at the DTI, Kenneth Clarke, announced to Parliament that BS had been approached about the disposal of Govan and Appledore, as well as contractual difficulties with the ferry programme at NESL. Just who approached whom, and when, remains obscure. To the new BS Chairman, John Lister, this came as especially bleak news in that, at least in his view, not only was the market situation improving but BS was better placed than at any time since nationalisation to take advantage of it. His main fear was that any more closures or disposals would take BS below critical mass and thereby limit its ability to take advantage of the rising market. Appearing before the Trade and Industry Committee in 1988, he was asked if he had made his views known to the DTI and, if so, what reply he had had. He answered that 'their response has been to suggest that we ought to sell Govan to Kvaerner'. Asked by the Committee Chairman if the implication was 'that your success means that you should sell off one of your yards', Lister's clearly depressed reply was, 'that is the situation I find myself in'. Further questions and answers were indicative of the position of the new Chairman relative to the Government:

Q: We must assume from this that you are 100 per cent opposed to selling Govan to this Norwegian company, are you?
A: I am under instruction. It does not matter whether I am 100 per cent opposed.
Q: So it is the Government . . . that is telling you to sell it?
A: That is right.
Q: Despite your business judgement which says that you should not sell it.
A: My judgement is that British Shipbuilders will have some difficulty in operating without this yard.
Q: So in practice your corporation would disappear?
A: Yes.[42]

Much of the questioning, however, missed the point of Clarke's earlier announcement, which had not been solely concerned with Govan. Although Lister was restrained in making the obvious point he certainly did so, admitting that 'now that we are down the track under instruction to sell off or to privatise as much of the remaining capacity as we can, we are liable to end up at the end of the year with NESL, nothing else'.[43] Despite the downbeat tone, it would prove a wildly optimistic statement. Lister was to be proved absolutely correct. On 27 June 1988 Kenneth Clarke announced the sale of Govan to Kvaerner for £6 million. But as with Scott Lithgow it was a curious deal, with 500 jobs lost immediately, BS picking up the redundancy costs and the eventual cost of the sale to BS being £19 million. Kvaerner also gained access to the SIF, which as a Norwegian company outwith the EU it had not been previously able to do—an irony in itself for SIF monies had been denied to all BS yards in 1988. The use of EU policy was becoming obvious.[44] By the end of 1988, Appledore was sold to Langham Industries at a net cost to BS of £3.4 million while Ferguson's was sold to the HLD Group at a cost to BS of £3.8 million.[45] Lister's prediction had been realised: BS was left with NESL, but not for long.

On 7 December 1998 the Chancellor of the Duchy of Lancaster, Tony Newton, announced that no bids had been received for NESL that did not involve the continuation of subsidy. Therefore, the yard would close. Newton's view was that the yard had lost £100 million and that if the money had been spent 'on providing alternative employment rather than on propping up this industry to make losses we should now be in a better position'.[46] The situation on the Wear, however, would reveal two separate issues: the Government's desire to end subsidy to BS, and the use of EU policy to drive this aim. Already compromised by the Danish ferry order, a situation which was no fault of NESL or BS, the yard was seeking an order from Cuba which merely awaited agreement on a financial package. Newton's predecessor, however, Kenneth Clarke, had decided that NESL had to be privatised before he would consider granting SIF assistance to the Cuban order. The instructions were clear: 'we shall not give intervention fund support to British Shipbuilders, because its history is that it cannot comply with the sixth directive and it cannot stop making cost overruns on top'.[47] Assistance would therefore be given to the private sector but not to the public sector. The issue then turned on whether a 'suitable buyer' for the yard could be found before the announcement of closure. The problem of finding a buyer for NESL was complicated by the fact that in December 1988 the Government had accepted EU counterpart funding to establish an enterprise zone in Sunderland. (The same affliction had been visited on other areas without their full realisation of what was going on.) As the EU Commissioner, Sir Leon Brittain, later made clear, a range of aid measures had been authorised and 'the closure of NESL formed part of the counterpart for that aid'.[48] Reduction of capacity was thus the *quid pro quo* for such aid, and this meant that the EU would take a dim view of any bids for the yards after 1988.

As it transpired, the package agreed in December 1988 included the sale of Appledore, Govan, Ferguson and Clark Kincaid to the private sector, the 'total closure' of NESL and the winding up of BS as a publicly owned company. Not only did the deal include restructuring aid, but also loss compensation for BS for 1987–89, of which NESL represented a major proportion because of the Danish ferries contract. The 1988 losses breached the sixth directive on subsidy ceiling and forced BS to borrow £90 million, although the Commission decided to average the 1987–89 losses because NESL was to close.[49] If the closure of NESL was not 'final and irreversible', it was likely that the Commission would demand repayment, re-examine Sunderland's status as an enterprise zone and force any new owner to shoulder the £90 million of cost overruns inherited from BS.[50] Having denied SIF support for the Cuban order, the Government then informed prospective bidders that they too would be denied access to the SIF. When a bid was finally received that was not contingent upon SIF support, the bidder was promptly told that it would have to inherit the £90 million in losses. That closure was pre-determined certainly seems to have been the view of the BS Chairman, John Lister. As Bob Clay, the Sunderland North MP tells it, Lister, at a meeting in Sunderland on 25 May 1988, declared that 'there is a bloody great steamroller moving towards Sunderland. I don't like it but there may not be much I can do about it.' Later the same day, he told Clay that 'You remember that steamroller I was talking about this morning, there is nothing whatever that you or I can do to stop it'.[51] With the closure of NESL, BS effectively disappeared. The overarching EU policy of reducing capacity in line with demand had been pursued with a fervour in Britain to the virtual elimination of serious mercantile capacity. As John Lister put it in 1988:

> we should be saying the same as the other Western European nations, that we have contracted dramatically, we need a merchant shipbuilding capacity. We have only 13 per cent of the merchant shipbuilding capability we had ten years ago so we have reduced more than anybody else.[52]

With the closure of NESL, 13 per cent became 6 per cent, and as BS disappeared, so too did volume shipbuilding in the UK.

With BS privatised, the Government decided, somewhat audaciously, to privatise Harland and Wolff. The yard had been 'saved' by Government intervention in the inter-war period, but since the onset of the economic crisis in 1975 it had consumed £532.7 million in public money down to 1988. As the Secretary of State for Northern Ireland, Tom King, made clear, there was no future for Harland and Wolff in the public sector, stressing that 'it . . . [was] a matter of urgency that we achieve a private sector involvement as a solution for Harland and Wolff'.[53] The yard's performance in the formal public sector had been, as Table 40 shows, truly gruesome. In King's view, Harland and Wolff, as the last remaining yard in public ownership, could not claim that it was competing on a fair basis with those companies which were exposed to the rigours of the market. As King asserted, it was 'absolutely clear that the situation could no longer continue in which there was one publicly owned . . . [yard] trying to maintain that it was bidding on a competitive and fair basis'.[54] All of which rather beggared belief as to the placement of the AOR vessel for the MoD.

The privatisation of Harland and Wolff was occasioned in mid-1988

Plate 10: Modernisation realised: the Doxford ship factory completed.

Table 40: Harland and Wolff: losses (profits) on ordinary activities per employee under public ownership

Date	Employees	Loss (profit) on ordinary activities (£000s)	Loss (profit) per employee (£)
1975	9,657	4,884	505
1976	9,236	(2,591)	(280)
1977	8,706	1,907	219
1978	8,212	25,452	3,099
1979	7,542	43,296	5,740
1980 (to 31.12)	7,370	31,997	4,341
1982 (to 31.3, 15 months)	7,034	26,205	3,725
1983	6,162	42,822	6,949
1984	5,452	29,756	5,457
1985	5,163	35,819	6,937
1986	4,937	28,988	5,870
1987	4,899	57,866	11,811
1988	4,044	17,269	4,270

Source: (1988–89) Trade and Industry Committee: The Privatisation of Harland and Wolff, HC 131, p. x.

when the Department of Economic Development of Northern Ireland began negotiations to sell Harland and Wolff to Tikko Cruise Line Ltd, with whom the company had been in negotiation for the construction of the P3000, the 'Ultimate Dream' cruise liner. Whilst the Government was prepared to consider intervention fund support for the contract, it was unwilling to accept payment on cost overruns and thus decided that the yard should be returned to the private sector. It was therefore suggested to the Tikko Chairman, Ravi Tikko, that he should acquire the yard in order to build the P3000. The negotiations, however, broke down, as did bids from Bulk Transport Ltd and UM Holdings. By October 1988, the favoured bid was a management/employee buy-out, with the Government willing to establish Harland and Wolff 'on the most generous basis that we can that establishes them in the private sector or any other ownership'. But it would not sustain 'the continuing liability and commitment of public funds for an unspecified period'. Still, given that no publicly owned yard was receiving SIF support, the decision to extend SIF support to Harland and Wolff was clearly anomalous.[55] Why Harland and Wolff as a mercantile yard was allowed to bid for MoD contracts is unclear, but it was not the only merchant yard to be placed in such a privileged position. It was though, the last remaining yard in the UK capable of building all types of surface vessels and was still the largest in the UK, having been extensively updated in the 1970s, and had the capacity to build up to 1 mgrt. The Trade and Industry Committee recommended the continuation of SIF support, but none for cost overruns, and whilst support for privatisation was firm, the Government had to support the yard on the basis 'that the company . . . [was] not put out of business as a result of foreign competition at a more highly subsidised level'.[56] This was an interesting view, in that assistance was extended to the Belfast yard, whilst being denied to yards on the mainland.

The paucity, however, of spraying naval orders around for political reasons was revealed through the process of the Harland and Wolff privatisation. The AOR order, for example, *Fort Victoria*, had turned into yet another financial disaster. As was later conceded, because Harland and Wolff had not maintained a warshipbuilding capability it had 'had considerable difficulties in discharging its obligations under this contract'. There had also been 'extensive delays both in delivery of design information and delivery of the vessel'. The result was that the Government paid the new owners of Harland and Wolff—a Norwegian consortium headed by Fred Olsen—£53 million in compensation and the vessel was completed at Cammell Laird.[57] It was also admitted that 'VSEL had won the LPH order by bidding at a loss', a fact earlier denied by both the CPA and NAO, and that if acquired by BAe which then put in a bid for Type 23 frigates 'by again bidding at a loss (cross-subsidised by its monopoly activities) there would be no likely alternative to the closure of the YSL (Yarrow) yard'.[58] Why it was, therefore, that no concern was expressed when BAe acquired GEC-Marconi Marine seems inexplicable. Despite the long mantra from the MoD over competitive tendering in warshipbuilding, the Government had ended up by allowing a near monopoly of major warship construction by 2000 (with BAe owning Yarrow, VSEL and Govan, and the only alternative being Vosper Thornycroft, which was limited by spatial constraints). In this equation, politicians and civil servants were equally culpable, in that the publicly stated policy of facilitating competition had actually produced a quasi-monopoly. The whole process had been ideologically driven, and whilst it was open to argument that the same process had applied to the 1974–77 Labour Government's policy, it could equally be argued that those Governments had attempted to 'save' the industry whilst the Conservative Governments were simply trying to divest.

The result of the policy of privatisation was the ultimate emasculation of the British shipbuilding industry as a major builder in world markets. The remaining merchant yards in 2001 were: Govan (BAe-owned), Harland and Wolff, Appledore and Ferguson's (with Ailsa closing in 2001). In terms of warship yards, the names remained familiar although the ownership did not: Yarrow and VSEL (both owned by BAe) and Vosper Thornycroft. The result of privatisation had been an income to BS of £125.5 million while costs had been £234.8 million, leaving a negative balance of £109.3 million.[59] Minister after minister at the DTI, and there were many of them, consistently stressed how much money BS had lost. Kenneth Clarke would claim invariably that the corporation had lost £1.8 billion in the ten years of its existence; indeed, the headline figure by 1989 was over £2 billion. When these figures were disaggregrated, however, it transpired that BS had paid out £313 million on redundancies, £228 million on Scott Lithgow, £93 million on customer failures such as International Maritime Fruit Carriers, and the Danish ferry contracts and £31 million on temporary borrowings and costs arising from privatisation. Capital investment accounted for £640 million and SIF support amounted to £258 million. What amounted to losses, therefore, came to £526 million, 60 per cent of which was accounted for by under-recoveries due to gaps in the building programmes and building losses.[60] This, however, took no account of the costs of redundancy and unemployment.

In an industry where boom and bust was fairly endemic, the redundancies of the 1980s were of a different scale from those which had occurred earlier. Between 1951 and 1971, the total number employed in the shipbuilding industry had declined by over one-third. This, however, was in a period of rapidly expanding job opportunities, something which could hardly be said to have characterised the 1980s and early 1990s. Moving jobs had been easy in the 1960s and early 1970s, but in the 1980s, with the institution of the redundancy scheme, jobs were effectively being sold. For example, BS estimated that the direct costs of making 3,500 people redundant in 1986 was between £35 and £45 million.[61] This, though, calculated only the direct costs. BS estimated an unemployment multiplier of between 1.5 and 3.1, whereas the DTI operated an unemployment multiplier of 1.7. It also argued that the validity in calculating the overall effect on employment of a given number of shipbuilding redundancies depended on the key assumption that shipbuilding output, and therefore the input from suppliers, would fall in proportion to the number of jobs lost. In 1988 Lister argued that there were basically two groups which benefited from shipyard employment: those working to support the completion of contracts and those benefiting in the local economy from the services required in the community. According to this analysis, up to 60 per cent of a typical ship's costs comprised materials, and Lister believed that the average direct multiplier was at least 1.5. Additionally, there was also the indirect employment generated in the local community via purchasing power and services where the multiplier was calculated at 2.[62] As closure and redundancy became the familiar pattern of the 1980s and 1990s, the hidden costs became all too apparent. In 1990 VSEL had employed nearly 50 per cent of the Barrow travel-to-work area, but the yard had shed some 5,000 jobs in two years and by June 1992 there were 251 vacancies for 4,300 people registered as unemployed. On the Wirral, the Cammell Laird closure had a catastrophic effect. In 1992 unemployment had reached 14 per cent, with some much more heavily concentrated clusters of idleness. Over 600 local firms had supplied the yard with £30 million of goods and services on an annual basis and the local authority calculated the cost in lost tax revenues, increased welfare payments and retraining at £111 million.[63]

It had been a spectacular and rapid denouement. From the inception of State support via the Shipbuilding Credit Scheme in 1963, it had taken less than 30 years for the British shipbuilding industry to disappear as a volume producer of ships. Paradoxically, the period of its elimination coincided with levels of support which would have astonished the inter-war leaders of the industry. Shipbuilding had been reorganised under the aegis of the SIB following the Geddes Report. By this stage, however, the market conditions had become torrid and the easy profits of the 1950s had given way to brusque competition and fixed-price contracts which inflation often rendered unremunerative. It may be the case that intervention via the Credit Scheme and the SIB merely allowed the shipbuilders to relax in the face of foreign competition. Certainly such modernisation as the industry did undertake was too little and too late, old markets had already been lost and new markets missed. The impact of the inter-war depression, whereby the industry had survived the slump through the use of labour-intensive methods and totally outmoded technology, had burned deep into the psyche. Some had recognised this as early as 1945. E.A. Seal, the Deputy Secretary of the Admiralty, for example, noted to the Director of Naval Construction:

> The principal weaknesses seem to me to be a lack of up-to-date equipment, and worse than that, a general failure on the part of the managements to recognise their deficiencies in this respect, or if they do recognise them a general disinclination to remedy them . . . I feel that the attitude of the managements is due first of all to the fact that they are nearly all old men and, even if young men, are steeped in the traditions of pre 1914 . . . even if they do recognise the necessity for more equipment they are reluctant to spend money on it because they know it is only when demand is high and steady that expensive equipment can justify itself financially . . . they are no doubt anxious for industrial peace and new methods mean labour trouble . . . The real fact of the matter is that instead of worrying about the future efficiency of the industry they are like a lot of small-minded pettifoggers simply worrying about a minor and literally insignificant detail and using their alleged lack of capital as an excuse.[64]

Seal went so far as to suggest a full-scale Committee of Inquiry into the efficiency of management, the adequacy of equipment in terms of meeting post-war competition, and whether the industry did, in fact, have adequate financial reserves. It was prescient, accurate and ignored.

By and large, Charles Lillicrap, the DNC, agreed with Seal's critique, commenting that what had exasperated him was that the 'Shipbuilders did not welcome with open arms the opportunity for capital development necessary to bring themselves up to date', with the only explanation being the 'innate conservatism and reluctance to face up to the fact that the old methods are outdated'. Reflecting on the condition of the industry in the wake of the inter-war years, he observed that it was 'a matter of wonderment that they have ever have [sic] produced ships at all'. Commenting on the impact of the inter-war slump, Lillicrap noted that 'the policy adopted has been one of pinch-back economy, whereas what was wanted was a bold facing up to modernisation'. The DNC, however, was the first to put the block on Seal's more radical proposals, declaring that he had 'serious doubts' about Seal's advocacy of an inquiry into management, technology and reserves on the basis that the 'industry would resent it and the opposition would be tremendous'.[65] This set in train a process whereby the industry would be a source of continued consternation in the 1940s and 1950s without any Government intervention. As such shipbuilding soldiered on, keeping the Government at arms length in the 1950s, but as its position deteriorated it made increasingly more desperate appeals to Government for assistance. The fact that Seal and Lillicrap's views in 1945 could stand as valid as late as 1965 and 1972 is no more than an indictment of the leadership of the industry. The Credit Scheme would be followed by Geddes and then the SIB, in turn by Booz-Allen and Hamilton, nationalisation to save the industry and then privatisation to get rid of it. It all availed nothing. By the late 1980s—the criticisms of 1945 could still resonate—the British shipbuilding industry had failed to meet the challenge of foreign competition and had been reduced to niche markets in international shipbuilding. Despite the consumption of massive resources from the state between 1962 and 1990 the industry had all but disappeared.

Epilogue

In September 2001 shipyard workers at the BAe-owned Clyde yards at Govan and Scotstoun held a one-day strike to protest against redundancies. In stark contrast to earlier years, when a shipbuilding strike would have generated national headlines, the stoppage occasioned little comment outside of Clydeside itself. The context was all too familiar: a Government committed to shipbuilding, the largest naval programme in living memory, but an industry reduced to rump status in international terms and struggling for its very existence.

The 1998 Strategic Defence Review proclaimed, however, in detail which did capture the imagination of newspaper headline writers, that the UK was to build two large aircraft carriers to be in service by 2012.[1] One year on from the Review, the Minister for Defence Procurement, Baroness Symons, announced that contracts worth £100 million would be placed for initial design assessments of the aircraft carriers, a new fleet of Type 45 destroyers to replace the ageing Type 42, and a number of fleet auxiliary vessels, in particular two alternative landing ship logistics vessels. The Government also indicated that it favoured spreading the bidding opportunities around the established warship yards. These were the BAe-owned yards at Scotstoun and Barrow, Vosper Thornycroft on the south coast, the merchant yards at Harland and Wolff and the BAe-owned Govan yard on the Clyde, Swan Hunter and Cammell Laird (which had been saved for ship conversion and repair), and the privatised former naval dockyards such as Babcock Marine at Rosyth on the Firth of Forth.

According to the Government, the total value of the new orders could reach £8 billion, with Baroness Symons describing the proposed naval programme as 'the biggest for a quarter of a century'.[2] A few months later, the Minister of State at the Scottish Office, Brian Wilson, added his support, declaring shipbuilding to be a 'modern industry of the future' whilst pledging assistance to its mercantile sector.[3] The outcome, however, was depressingly familiar. Despite Wilson's pledges, the Ailsa yard at Troon could not close the gaps in its order book and therefore shut down in the winter of 2000. This left Govan, Scotstoun and Ferguson's at Port Glasgow as the only yards functioning on the Clyde. Nor did the bold pronouncements on the naval programme come soon enough for Cammell Laird, which was operating repair and conversion yards on the Mersey, Tyne and Wear, and had risen almost phoenix-like from the ashes after VSEL had disposed of it in the 1980s. It had established a strong presence in the repair/conversion market and by the end of 2000 employed over 1,800 people, while its shares were trading at £1 each. The company not only had a contract with the Italian company Carnival to convert a cruise liner, *Costa Classica*, worth £40 million, but was also in the market for some of the six naval support vessels, and was awaiting a package of DTI assistance for the construction of two cruise liners for Luxus. In the event, the six MoD orders were split 2:4 between Harland and Wolff and a German yard, the *Costa Classica* turned around halfway between Genoa and Birkenhead, and the DTI failed to produce an aid package which would satisfy both Cammell Laird and Luxus. Cammell Laird duly defaulted on its overdraft, the shares slumped to 6p and by April 2001 the receivers were in.[4] Negotiations continued throughout the summer and autumn of 2001 as attempts were made to save the yard with the ship-repairers and offshore contractors A&P, the favoured buyers. The firm's future, however, remains fundamentally uncertain.

The Type 45 destroyer orders did provide some relief, but not enough to prevent BAe announcing over 1,000 redundancies at Govan, Scotstoun and Barrow. Once again, the company blamed the MoD for a failure to bring forward interim work on auxiliary vessels.[5] In the wake of the redundancy announcement, the Scottish Executive established a Clyde Shipyards Taskforce to examine (yet again!) the viability of shipbuilding on the Clyde and the alternative employment and training prospects.[6] At Swan Hunter, the news was slightly better. The yard had been purchased by the Dutch businessman Jaap Kroese in 1995 from under the noses of the receivers, and now concentrated on the conversion of ships into pipe-laying vessels and production platform related work. In October 2001, however, the yard began cutting tons of steel for two of the Royal Navy's alternative landing ship logistics vessels. The company expects to increase its workforce from 500 in 2001 to around 2,000 by 2004 and intends to start 100 apprentices a year for the foreseeable future. As the Swan Hunter yard services manager, Brian Gardner, views the position, 'far from being a sunset industry, we can see huge opportunities in the future and we are bidding for a lot of work.

This country might not build many ships any more, but others in Europe (Finland, Germany, France, Italy) do and there is a market to be tapped.'[7]

It all remained to be seen, however. Would the naval programme remain at its current level? In particular, would the much trumpeted aircraft carriers be built? Extrapolating even further, would MoD willingness to order from yards bring private mercantile orders in their wake? Given the modern history of British shipbuilding there could be no guarantee. By 2001, few yards remained. On the south coast, Vosper Thornycroft announced its intention to move from Southampton to the site of the former Portsmouth naval dockyard. In Devon, Appledore looked secure in its highly specialised niche market for small ships. At Belfast, Harland and Wolff limped along, sustained as always by the peculiar politics of Northern Ireland. In England, the BAe yard at Barrow remained beholden to the submarine market although it was desperately trying to re-enter the surface ship market. Cammell Laird appeared doomed whilst Swan Hunter appeared on the rise, but still, the most modern ship factory in Europe, at Pallion on the Wear, lay in mothballs despite various attempts to reopen it. On the Clyde, outwith the two Bae-owned yards on the Upper Reaches, only Ferguson's at Port Glasgow survives. Now, as little more than bit players in international markets, it would be a bold Briton who could assert seriously that 'the nations' were 'not so blest as thee' and that Britain remained 'the dread and envy of them all'.

Notes

Introduction

1. What follows is a general political economy of the British shipbuilding industry in the twentieth century. As such, the constraints of space do not permit full consideration of issues which are worthy in and of themselves of single-volume studies. A comparatively short list would include: the performance of the industry in both the First and Second World Wars, industrial relations, technology and the regional variations in performance. Full consideration could also be given to the national and regional employers' associations and their interactions.
2. M. Davies, *Belief In the Sea* (1992), pp. 9–54.
3. See A. Slaven, 'British Shipbuilders: Market Trends and Order Book Patterns between the Wars', *Journal of Transport History*, 3 (1982), pp. 37–61.
4. A. McKinlay, 'The Interwar Depression and the Effort Bargain: Shipyard Riveters and the Workers Foreman, 1919–1939', *Scottish Economic and Social History*, 9 (1989) pp. 55–70.
5. For comparisons with the shipbuilding industry, see M.W. Kirby, *The British Coalmining Industry 1870–1946* (1977) and S. Tolliday, *Business, Banking, and Politics: The Case of British Steel, 1918–1939* (Cambridge, Mass., 1987).

1 Sea Change

1. Public Record Office (hereafter PRO), RECO1/276, 'Future Status of the Mercantile Marine', July 1919.
2. PRO, RECO1/274, 'Committee on Commercial and Industrial Policy to the Reconstruction Committee', June–August 1917.
3. *Lloyd's Register of Shipping*, various years.
4. PRO, BT55/113, 'Interim Report of the Departmental Committee Appointed by the Board of Trade to Consider the Position of the Shipping and Shipbuilding Industries after the War', 1917.
5. PRO, RECO1/279, Letter from the British Embassy, Tokyo, to the Secretary of State for Foreign Affairs (A.J. Balfour), 9 August 1917.
6. PRO, RECO1/321, Memorandum Regarding the Danish Shipbuilding and Repairing Yards, November 1917.
7. PRO, RECO1/282, Conference between Ministers and Shipbuilders and Shipowners at the Ministry of Shipping, 14 March 1918.
8. PRO, RECO1/282, Report on the Shipbuilding Yards, August–September

1918.

9. PRO, RECO1/282/3754A, 'Control of Shipping After the War', undated but 1918. See also RECO1/271, 'Shipping Facilities After the War', March 1917, and RECO1/276, 'Merchant Shipbuilding: Policy of Government. Shipowners Anxiety over the Possibility of Nationalisation', 1917.
10. Quoted in J.F. Clarke, *Building Ships on the North East Coast: Vol. 2, c. 1914–1980* (Whitley Bay, 1997) p. 23.
11. On the issue of bespoke markets and the implications thereof, see A. Slaven, 'British Shipbuilders: Market Trends and Order Book Patterns between the Wars', *Journal of Transport History*, 3 (1982), pp. 37–61.
12. Quoted in H.B. Peebles, *Warshipbuilding on the Clyde* (Edinburgh, 1987) p. 96.
13. Clarke, *Building Ships*, pp. 225 and 238.
14. M. Moss and J.R. Hume, *Shipbuilders to the World: 125 Years of Harland and Wolff, Belfast 1861–1986* (Belfast, 1986), pp. 208–13.
15. *Shipbuilding and Shipping Record*, 1 May 1916.
16. *Shipbuilding and Shipping Record*, 1 April 1920.
17. *Shipbuilding and Shipping Record*, 3 June 1920 and 22 July 1920.
18. *Shipbuilding and Shipping Record*, 2 September 1920.
19. *Shipbuilding and Shipping Record*, 9 and 23 December 1920.
20. Moss and Hume, *Shipbuilders*, p. 231.
21. Peebles, *Warshipbuilding*, pp. 97–8.
22. Clarke, *Building Ships*, pp. 265–6.
23. Moss and Hume, *Shipbuilders*, pp. 209–10.
24. *Shipbuilding and Shipping Record*, 5 May 1921.
25. *Shipbuilding and Shipping Record*, 18 August 1921.
26. A. Slaven, 'A Shipyard in Depression: John Brown's of Clydebank, 1919–38', *Business History*, 19 (1977), p. 126.
27. *Shipbuilding and Shipping Record*, 23 March 1922.
28. Slaven, 'A Shipyard', p. 127.
29. *Shipbuilding and Shipping Record*, 20 October 1921.
30. PRO, CAB24/128 (CP3363), 'Proposals of Hilton Young Presented to the Prime Minister at Gairloch', 2 November 1921.
31. PRO, CAB27/120 (CU274), 'Programme to Relieve Unemployment' 11 October 1921.
32. The records of the Advisory Committee are contained under various classifications at the PRO. See CAB 24/136 (CP 3907); T190/1–276; T160/783 F 13861/1–3 and T172/1208 (1); and also T160/241/F9159 and 244/F9385, 'Trade Facilities Act, 1925–1929'.
33. *Shipbuilding and Shipping Record*, 20 April 1922.
34. *Shipbuilding and Shipping Record*, 18 May 1922.
35. *Shipbuilding and Shipping Record*, 3 January 1923.
36. *Shipbuilding and Shipping Record*, 13 July 1922.
37. J.E. Mortimer, *History of the Boilermakers' Society: Vol. 2, 1906–1939* (1982), pp. 139–44.
38. PRO, CAB27/197 (CU712), Unemployment Committee, Unemployment Sub-Committee on Assistance to Shipbuilding, Report, 2 December 1924.
39. *Shipbuilding and Shipping Record*, 13 March 1924.

40. *Shipbuilding and Shipping Record*, 31 January and 24 April 1924.
41. PRO, CAB27/197 (CU 7120), Unemployment Committee, Unemployment Sub-Committee on Assistance to Shipbuilding, Report, 2 December 1924.
42. *Shipbuilding and Shipping Record*, 12 March 1925.
43. National Maritime Museum (hereafter NMM), SRNA 4/F1, Foreign Competition 1938–1951, Report of a Joint Inquiry into Foreign Competition in the Shipbuilding Industry, 1938.
44. *Shipbuilding and Shipping Record*, 2 April 1925.
45. *Shipbuilding and Shipping Record*, 9 April 1925.
46. Slaven, 'A Shipyard', p. 128.
47. Peebles, *Warshipbuilding*, p. 115.
48. *Shipbuilding and Shipping Record*, 30 December 1926.
49. J.R. Parkinson, 'Shipbuilding' in N.K. Buxton and D.H. Aldcroft (eds), *British Industry between the Wars* (1979), p. 80.
50. For details, see L.A. Ritchie, *The Shipbuilding Industry: A Guide to Historical Records* (Manchester, 1992), *passim*.
51. Sperling, through Northumberland, owned the Furness interest at Irvine, all of the share capital of Doxford, 85 per cent of the share capital of Fairfield, the entire share capital of the Monmouth Shipbuilding Company and of Workman Clark, and a controlling interest in Blythswood on the Clyde. See J.R. Parkinson, *The Economics of Shipbuilding in the United Kingdom* (Cambridge, 1960), pp. 34–5.
52. A. Armitage, 'Shipbuilding at Belfast: Workman Clark and Company, 1880–1935' in L.R. Fischer (ed.), *From Wheel House to Counting House: Essays in Maritime Business History in Honour of Professor Peter Neville Davies* (St John's, 1992), pp. 113–22; Peebles, *Warshipbuilding*, pp. 124–8; and *Fairplay*, September 1925.
53. PRO, T172/1670, 'Extension of Trade Facilities', Memorandum by Sir Richard Hopkins, 17 October 1929.
54. *Shipbuilding and Shipping Record*, 1 March 1928.
55. NMM, SRNA 4/S50, Papers Prior to Formation; and Proposals for a Tendering Expenses Scheme (and Other Methods to Aid Shipbuilding) 1926–30.
56. PRO, BT56/28, Report of a Meeting with the Lord Privy Seal, the Rt. Hon. J.H. Thomas, at the Treasury, London, 8 November 1929.
57. *Glasgow Herald*, Annual Supplement, 31 December 1929.
58. PRO, BT56/28, 'Survey of Shipbuilding Tonnage Position and Prospects At the End of March 1930', Memorandum by the Shipbuilding Employers' Federation, 25 April 1930.
59. NMM, SRNA 7, SC1/1, Shipbuilding Conference, Minutes of Private Meetings, 1928–1933, and A. Slaven, 'Self-Liquidation: The National Shipbuilding Security Ltd and British Shipbuilders in the 1930s' in S. Palmer, and G. Williams (eds), *Charted and Uncharted Waters: Proceedings of a Conference on the Study of Maritime History* (1982).

2 The Weight of History

1. R.H. Campbell, 'Costs and Contracts: Lessons from Clyde Shipbuilding Between the Wars' in A. Slaven and D.H. Aldcroft (eds), *Business, Banking and Urban History: Essays in Honour of S.G. Checkland* (Edinburgh, 1982) p. 63.
2. On the activities of NSS, see J.R. Parkinson, 'Shipbuilding' in N.K. Buxton and D.H. Aldcroft (eds), *British Industry Between the Wars* (1979), pp. 96–8, L. Jones, *Shipbuilding in Britain: Mainly Between the Two World Wars* (Cardiff, 1957), pp.133–40 and A. Slaven, 'Self-Liquidation: The National Shipbuilding Security Ltd and British Shipbuilders in the 1930s' in S. Palmer and G. Williams (eds), *Chartered and Unchartered Waters: Proceedings of a Conference on the Study of Maritime History* (1982), pp. 125–41.
3. Quoted in L. Johnman, 'The Shipbuilding Industry' in H. Mercer, N. Rollings and J. Tomlinson (eds), *Labour Governments and Private Industry* (Edinburgh, 1992). p. 187.
4. Bank of England, Securities Management Trust Papers (SMT), SMT2/280, National Shipbuilders Security Company Limited, 24 April 1929. For a discussion of the rationalisation movement, see L. Hannah, *The Rise of the Corporate Economy* (1983), pp. 27–40.
5. Bank of England, SMT 8/15, Shipbuilding Employers' Federation Report, 21 May 1929.
6. PRO, BT 56/28/CIA/1381/1, Shipbuilding Industry's Scheme for Elimination of Redundant Capacity, 28 February 1930.
7. *Shipbuilding and Shipping Record*, 6 March 1930.
8. PRO, BT 56/28/CIA/1381/1, Letter from the Shipbuilding Employers' Federation to the Chief Industrial Advisor, Sir Horace Wilson, 2 May 1930.
9. H.B. Peebles, *Warshipbuilding on the Clyde* (Edinburgh, 1987), pp. 121–32, and G.A.H. Gordon, *British Seapower and Procurement between the Wars* (1988) pp. 76–91. The situation was so bad that the Warship Group of the Shipbuilding Conference advanced the following offer: 'In view of the grave situation for the industry a deputation met the First Lord of the Admiralty in December 1930 (the Minister of Labour, the Controller of the Navy and a Director of Warship Production also attended) and explained the serious plight of the industry due to the almost complete absence of demand. The industry's representatives referred in particular to the lack of new warship work and offered to build warships on extended credit terms, in order to help both the government and the industry. The offer, however, was not accepted.' NMM SRNA4 C11/1, 'Shipbuilding in the UK and Events Leading to the Formation of the Shipbuilding Conference, Offer to Build Warships on Credit Terms', 1930.
10. J.R. Hume and M.S. Moss, *Beardmore: The History of a Scottish Industrial Giant* (1979), p. 216, and I. Johnston, *Beardmore Built: The Rise and Fall of a Clydeside Shipyard* (Clydebank, 1993) pp. 100–51.
11. For an analysis of the formation and subsequent activities of NSS, see Slaven, 'Self-Liquidation', pp. 125–41.
12. *Transactions of the Institution of Engineers and Shipbuilders in Scotland, Presidential Address by Sir James Lithgow* (October 1930).

13. PRO, BT 56/28/CIA/1381/1, Shipbuilding Facts and Figures. Letter from Shipbuilding Employers' Federation to the Board of Trade, December 1930.
14. A. Slaven, 'A Shipyard in Depression: John Brown's of Clydebank, 1919–38', *Business History*, 19 (1977), p. 131. For an excellent history of John Brown's see I. Johnston, *Ships for a Nation, 1847–1971: John Brown and Company, Clydebank* (West Dumbartonshire, 2000). For the inter-war period, see pp. 164–213.
15. F.E. Hyde, *Cunard and the North Atlantic, 1870–1973* (1975), p. 203.
16. PRO, T161/252/S27224, 'Cunard Steamship Company'. Correspondence on renewing of Government agreement covering loans, terms for the use of Cunard ships during war, and mail contracts, 1925–1927.
17. Hyde, *Cunard*, p. 204.
18. Sir Percy Bates, quoted in Hyde, *Cunard*, p. 204.
19. PRO, T161/281/F11798/1, 'Cunard Request for Government Assistance to Build One or Two Large Passenger Liners', 28 January 1930.
20. PRO, T161/281/F11798/1, 5 May 1930, 21 May 1930 and 23 June 1930.
21. Cunard (Insurance) Agreement Act, 21 Geo.5c2.
22. E. Green and M. Moss, *A Business of National Importance: The Royal Mail Shipping Group, 1902–1939* (1982), *passim*. See also M. Moss. and J.R. Hume, *Shipbuilders to the World: 125 Years of Harland and Wolff, Belfast 1861–1986* (Belfast, 1986), pp. 279–322.
23. Green and Moss, *A Business*.
24. Churchill College, Cambridge, Weir Papers, Weir 11/1, Accounts of Earlier Negotiations: Cunard–White Star Negotiations During Year 1930.
25. PRO, T190/79, 'Cunard-Oceanic Merger', Memorandum from E. St J. Bamford to Sir R. Hopkins, 8 January 1931.
26. PRO, T190/79, Meeting at the Treasury, 6 July 1931.
27. Weir Papers, Weir 11/8, Trading and Financial Performance of Cunard, 1922–1931.
28. *Lloyd's List Annual Review*, 1931.
29. D. Kirkwood, *My Life of Revolt* (1935) p. 252.
30. PRO, MT9/2714 M5119, 'Suspension of Work on New Cunarder', letter from Sir Percy Bates to the Prime Minister, 15 February 1932.
31. PRO, T190/80, 'Cunard-Oceanic', Memorandum of a meeting at the Treasury, 18 February 1932; and MT 9/2714 M5119, Memorandum by Sir Charles Hipwood, 18 February 1932.
32. Green and Moss, *A Business*, p. 162.
33. PRO, T190/82, 'Cunard-Oceanic' unsigned and undated memorandum, but March 1934. The extract quoted is the genesis of a series of Treasury briefing papers for the Chancellor of the Exchequer attempting to defend the change of policy with regard to the North Atlantic.
34. PRO, T190/82, 'Cunard-Oceanic Merger', unsigned and undated Memorandum (March 1934).
35. *The Times*, 9 February 1934 and 14 December 1933. The figure of 105,000, of course, allowed for employment in suppliers and ancillary industries.
36. PRO, T190/83, 'Cunard-Oceanic Merger', Letter from Sir Warren Fisher to Sir Percy Bates, 3 May 1934; Note from Fisher to Neville Chamberlain and

Note from Chamberlain to Fisher, 3 May 1934.

37. Churchill College, Cambridge, Grigg Papers, Grigg 2/16–2/19, Letter from Montagu Norman to P.J. Grigg, 28 September 1934.
38. *Financial Times*, 21 November 1931.
39. *Shipbuilding and Shipping Record*, 26 March 1931.
40. *Shipbuilding and Shipping Record*, 16 July 1931.
41. *Shipbuilding and Shipping Record*, 19 November 1931.
42. *Glasgow Herald Trade Review*, 31 December 1931.
43. *Shipbuilding and Shipping Record*, 11 February 1932. See also *The Yorkshire Post Trade Review*, 14 January 1931, where the Vice-President of the Shipbuilding Employers' Federation and Managing Director of Smith's Dock, G. Tristram Edwards, claimed that the slump in shipbuilding was 'unparalleled', with only 10 per cent of berths occupied and that throughout 1931 90 per cent of firms had not booked a single order. Echoing Sir Frederick Lewis, Edwards claimed that 'no industry has in recent years suffered more from foreign competition of the most extreme kind, supported by State assistance in one form or another'.
44. E. Wilkinson, *The Town that was Murdered: The Life Story of Jarrow* (1939), *passim*.
45. PRO, MT9/2714, Letters from Sir John Ellerman to the Prime Minister, 16 February 1932, and to the Lord President of the Council, 12 September 1932.
46. *Shipbuilding and Shipping Record*, 12 May 1932.
47. PRO, CAB 27/557 CM (33) 2, 'Committee on the British Mercantile Marine', Recommendations of the Tramp Committee of the Chamber of Shipping, 9 December 1933.
48. PRO, CAB 27/557 BMM (33) 1, 'Committee on the British Mercantile Marine', Conclusions of the First Meeting, 18 December 1933.
49. PRO, CAB 27/557 BMM (33) 6, 'Committee on the British Mercantile Marine', Further Memoranda on the Present Position and Prospects of British Shipping, 22 December 1933.
50. PRO, CAB 27/557 BMM (33) 8, 'Committee on the British Mercantile Marine', Further Memoranda on Present Position and Prospects of British Shipping, 6 January 1934.
51. PRO, CAB 27/577, Letter from O.S. Thompson, Chairman, Aberdeen and Commonwealth Lines to W.R. Runciman, President of the Board of Trade, 13 February 1934.
52. PRO, CAB 27/577, Letter from the Chamber of Shipping and Liverpool Steam Ship Owners' Association to the President of the Board of Trade, 24 February 1934.
53. PRO, CAB 29/577 BMM(33) 13, 'Committee on the British Mercantile Marine', Summary of Answers to Questionnaire, 16 March 1934; CAB 37/557 BMM(33)14, Answers of Tramp Ship Owners, 24 March 1934; CAB 27/557 BMM(33)15, Comments On the Case Put Forward by Tramp Owners, 24 March 1934; CAB 27/577 BMM(33)16, Joint Declaration by the Chamber of Shipping and the Liverpool Steam Ship Owners' Association, 24 March 1934, and CAB 27/577 BMM(33)17, Note by the President of the Board of Trade, 26 March 1934.
54. PRO, CAB 27/577 BMM(33) 2, 'Committee on the British Mercantile

Marine', Conclusions of the Second Meeting, 27 March 1934.

55. PRO, CAB 27/577 BMM(33) 3, 'Committee on the British Mercantile Marine', Meeting, 24 April 1934.
56. PRO, CAB 27/577 BMM(33) 4, 'Committee on the British Mercantile Marine', Conclusions of the Fourth Meeting, 8 May 1934.
57. G. Coster, *Corsairville, The Lost Domain of the Flying Boat* (2000). In 1926 the Union Castle ship *Windsor Castle* left Cape Town for Southampton at the same time as a De Havilland bi-plane. Despite the fact that the aeroplane's route was over 60 per cent longer than the route of the ship, the aircraft landed two days before the ship docked (pp. 83–4).
58. PRO, CAB 27/577 BMM(33) 5, 'Committee on the British Mercantile Marine', Minutes of the Fifth Meeting, 11 June 1934.
59. The papers of the committees which administered Parts 1 and 2 of the Act are: PRO, BT14/11 and 155, the Tramp Shipping Committee 1935–1938; and PRO, MT9/2766 and 2767, the Ships Replacement Committee, 1935–1937.
60. Shipping (Assistance) Act 1935, Report of the Ships Replacement Committee, 1937, Cmd 5459.
61. Jones, *Shipbuilding*, pp.146–55.
62. PRO, CAB 26/19, Home Affairs Committee, 'British Shipping (Continuance of Subsidy) Bill', Memorandum by the President of the Board of Trade, 15 January 1936.
63. Sixth Report of the Tramp Shipping Advisory Committee, Cmd 5750, Vol. 13.
64. PRO, CAB 27/1656 S1(39)1, 'Committee on the Shipping Industry', Composition and Terms of Reference, 20 January 1939.
65. PRO, CAB 27/1656 S1(39)1, 'Committee on the Shipping Industry', Composition and Terms of Reference, 20 January 1939.
66. PRO, CAB 27/656 S1(39) 3, 'Cabinet Committee on Shipping and Shipbuilding', Proposals Regarding Tramp Shipping, Shipbuilding, Reserve of Tonnage and Liners, 10 March 1933.
67. PRO, CAB 27/656 S1(39) 2, 'Cabinet Committee on the Shipping Industry', Minutes of Second Meeting, 14 March 1939.
68. PRO, CAB 102/440, Merchant Shipbuilding and Repairs in the Second World War, draft for Official History by C.C. Wrigley.
69. PRO, CAB 102/4, British War Production, draft for Official History by M.M. Postan.
70. The Minutes of the Committee of Imperial Defence PSOC are at PRO, CAB 60, as are reports of the various sub-committees.
71. University of Glasgow Business Archives Centre (hereafter UGBAC), DC 35/31, Sir James Lithgow Papers, Committee of Imperial Defence, PSOC and Co-operation with Industry, Note of a Meeting, 19 December 1933. Lithgow served on the Advisory Panel with Lord Weir of Eastwood and Sir Arthur Balfour. Weir firmly believed that 'approaches should only be made to the big firms and the big men'.
72. In addition to Ayre and Lithgow, the other shipbuilders appointed to the SCC were Sir Maurice Denny of Denny Brothers, Sir Charles Craven of Vickers-

Armstrong, and F.C. Pyman of William Gray.

73. Gordon, *British Seapower*, pp. 172–246; and Peebles, *Warshipbuilding*, pp. 137–56.
74. W. Ashworth, *Contracts and Finance* (1952), pp. 106–9.
75. Parkinson, *Shipbuilding*, p. 97.

3 The Challenge of War

1. We concentrate here on the larger private establishments federated to the SEF, which represented the vast majority of firms in the industry. We do not consider in detail the contribution of the non-federated boatbuilding sector, the Admiralty-controlled Royal Dockyards, nor the war at sea. In the latter case the reader is directed to the standard work on the topic by S. Roskill, *The War at Sea*, 3 volumes (1954–61). For those readers seeking a more detailed appraisal of warship building during the war, see, I. Buxton, *Warship Building and Repair During the Second World War* (Glasgow, 1998). For merchant shipbuilding see also Sir Amos Ayre, 'Merchant Shipbuilding during the War', *Transactions of the Institute of Naval Architects* (April 1945). The authors have made extensive use of the material contained in the CAB 102/103 Series of records held at the Public Record Office, Kew, which contain preliminary draft material used for subsequent official histories of the war.
2. NMM, SRNA, Report of a Meeting of the Shipbuilding Conference, 13 October 1939.
3. UGBAC, DC 35/27, Sir James Lithgow Papers, Letter from Sir James Lithgow to Sir R.H.A. Carter at the Admiralty, 15 February 1940.
4. NMM, SRNA 7/SC/1, 1940, Report of a General Meeting of the Shipbuilding Conference.
5. For George Edwards's role as Director of Ship Repairs from June 1917 to January 1919, see J. Robinson, 'How Ship-Repairing Helped to Win the War', *Smith's Dock Monthly*, December 1919. For the 1939–45 War, see Sir L. Edwards, 'The War Effort and Organisation of British Shiprepairing', *Transactions of the North East Coast Institution of Shipbuilders and Engineers*, (1946–47). Sir James Lithgow had served as Director of Merchant Shipbuilding from 1917, on his return from an artillery unit in France, where he was responsible for the programme of standard ships, before returning to his Port Glasgow shipyards at the end of the war. Similarly, Sir Amos Ayre, who with his brother, Wilfrid, owned the Burntisland shipyard in Fife, had also served as a District Controller in Scotland.
6. On this point, the view, subsequently repeated by other ministers, of the President of the Board of Trade, Walter Runciman, was plain: 'NSS is a private undertaking, and Government approval of its plans is not required'. House of Commons Official Report, Vol. 286, cols 1168–69, 20 February 1934.
7. Ayre, 'Merchant Shipbuilding', p. 3.
8. NMM, SRNA 4, 555 and 555/1, Report of an AGM of the Shipbuilding Conference, 4 November 1938. By this stage the Shipbuilding Consultative Committee had determined that allocation should be on the basis that firms

built ships that they were accustomed to build, and for which plans and equipment etc. were available.

9. *Shipping World*, 9 January 1946. The wartime output of Lithgows Limited from 3 September 1939 to 31 December 1945 comprised 84 vessels of 540,000 grt. Doxford completed 80 vessels of 544,000 grt and William Gray completed 90 vessels of 408,030 grt.
10. The 'Big Six' Warship Group firms comprised Fairfield and John Brown on the upper Clyde, Swan Hunter on the Tyne, Vickers Armstrong at Barrow-in-Furness and Walker on the Tyne, Cammell Laird on the Mersey, and Harland and Wolff in Belfast. W. Hornby, *Factories and Plant* (1957), p. 43. Other Warship Group members such as Hawthorn Leslie kept two slips for mercantile construction, as did Cammell Laird and Denny, whilst Swan Hunter, Stephen and Vickers Armstrong kept three each.
11. As well as the 'Big Six', the fifteen-member Warship Group comprised Denny at Dumbarton, Stephen at Linthouse, Yarrow at Scotstoun, and Scott's of Greenock on the Clyde; J. Samuel White at Cowes, and Thornycroft on the south coast district; and Hawthorn Leslie on the Tyne. All of these firms had their own engine works. In addition, two engine-building firms were also members. One, Wallsend Slipway, was a subsidiary of Swan Hunter, and normally sub-contracted hull construction to its parent company, and the other, Parsons Marine Steam Turbine, normally sub-contracted hull construction to Vickers Armstrong's Tyne yard.
12. Buxton, *Warship Building*, pp. 3–4.
13. Roskill, *The War*, Vol. I, p. 106.
14. Buxton, *Warship Building*, pp. 3–4
15. PRO, CAB 102/440, Merchant Shipbuilding and Repairs in the Second World War, by C.C. Wrigley, unpaginated.
16. PRO, CAB 102/407, Labour Requirements and Supply, Shipbuilding and Engineering 1939–1945, p. 40. Peggy Inman used this master copy in draft for her official history, *Labour in the Munitions Industries* (1957). In the opinion of Sir Amos Ayre, the inter-war depression had 'adversely affected' the quality of the workforce. Moreover 'the large proportion of workpeople in the older age groups' had, for some time before the war, been a problem. At the outbreak of war, the calling up of apprentice labour and territorial and naval reservists exacerbated this problem. Ayre, 'Merchant Shipbuilding', p. 4.
17. PRO, ADM 116/4892, Shipbuilding Contracts Placed with Shipyards in USA and Canada.
18. *Shipbuilding and Shipping Record*, 6 December 1945, p. 561.
19. Ayre, 'Merchant Shipbuilding', p. 12.
20. PRO, CAB 102/440, Merchant Shipping and Repairs, Vol. II, unpaginated.
21. PRO, CAB 102/4, British War Production by M.M. Postan, p. 60.
22. Buxton, *Warship Building*, p. 4.
23. PRO, CAB 102/4, p. 62.
24. Roskill, *The War*, Vol. III, p. 304, notes that in the 1914–18 war, enemy submarines sank a considerably larger amount of ships (4,837) than was the case in the 1939–45 war (2,828); but that the tonnage sunk was some three and a half million tons less (11,135,440 grt) in the First, than in the Second

(14,687,231 grt) World War.

25. Hornby, *Factories and Plant*, p. 69.
26. For a comprehensive treatment of the naval repair function, see G.A. Bassett, 'The Repair and Upkeep of HM Ships and Vessels in War', *Transactions of the Institute of Naval Architects*, 88 (1946).
27. Inman, *Labour*, p. 88.
28. House of Commons, Parliamentary Papers 1942–43 Vol. II, Report of the Committee of Public Accounts, 13 October 1943, paras 18–20.
29. PRO, CAB 102/407, p. 191.
30. Inman, *Labour*, p. 141.
31. PRO, LAB 10/363, DCIC Scotland, Weekly Reports, 20 June 1942.
32. *Shipbuilding and Shipping Record*, 25 July 1940, p. 86. By this stage regarding the point of reinstatement of former boilermakers, the Society's General Secretary, Mark Hodgson, had candidly admitted that 'few of his really employable members were out of work'.
33. J.E. Mortimer, *History of the Boilermakers' Society: Vol. 3, 1940–1989* (1994), p. 10.
34. PRO, CAB 102/4, p. 97.
35. *Shipbuilding and Shipping Record*, 27 March 1941, p. 290.
36. P. Summerfield, *Women Workers in the Second World War: Production and Patriarchy in Conflict* (1984) p. 151. Summerfield notes that in comparison to female participation rates of 60 per cent in electrical engineering and 34 per cent in general engineering, shipbuilding fared badly. However, Summerfield also noted that a rate of participation of 16 per cent was achieved in marine engineering. As many shipbuilding firms had their own engineering works, the numbers of females employed in the industry was in fact greater than official figures suggest if shipbuilding was taken in isolation. From an initial figure of 2 per cent in 1939, female employment in the private shipbuilding and repair firms peaked at 6.6 per cent of the workforce in August 1944.
37. PRO, CAB 102/407, p. 98.
38. NMM, SRNA/1 SEF/3044, the effect of Clause 8 of the National Agreement was to exclude the pre-war female workforce that had remained in the industry, notably French polishers, from its provisions. This gave rise to female diluted labour being awarded higher wages after a suitable period of training than those which pertained for the pre-war female workforce.
39. PRO, LAB 8/662, Efforts to Secure the More Efficient Use of Labour Supply in the Clydeside Area and in the Shipbuilding Industry, Shipyard Labour Supply, East of Scotland, 15 August 1942. At HM Royal Dockyard, Rosyth, 500 out of 892 women were employed in a productive capacity. This compared with 292 women who were contemporaneously employed at five private shipyards, Burntisland, Caledon, Grangemouth Dockyard, Hall Russell, and Henry Robb. For the national picture, see H. Murphy, 'From the Crinoline to the Boilersuit: Women Workers in British Shipbuilding during the Second World War' *Contemporary British History*, 13, 4 (1999), pp. 82–104.
40. PRO, CAB 102/440, Wrigley gives this number, which ensued as a result of the Industrial Registration Order (1941) No., but does not make the obser-

vation.

41. PRO, CAB 102/407, p. 105.
42. Emergency Laws (Transitional Provisions) Act 1946, Emergency Laws (Miscellaneous Provisions) Act 1947, Emergency Laws (Continuance) Order 1948 and 1949, Restoration of Pre War Practices Act 1950. Thereafter the Emergency Laws were extended by Orders in Council until the Emergency Laws Repeal Act 1959.
43. PRO, CAB 102/407, p. 105.
44. PRO, CAB 102/434, Allocation of Manpower, First Draft by A.V. Judges, May 1948, p. 131. Compulsion of the available pool of female labour was enshrined in the National Service (No. 2) Act, December 1941.
45. NMM, SRNA/7/SEF/4/104, Circular Letter No. 148/46, September 1946. The statutory authority requiring a National Service Officer's permission before a worker left or was discharged from an establishment covered by the Essential Work (Shipbuilding and Repairing) Order 1942 was dispensed with when the Order was withdrawn on 31 December 1946.
46. PRO, CAB 102/379, Industrial Relations and Welfare in Admiralty Establishments and Contractors Works, unpaginated draft.
47. PRO, CAB 102/19, *Manpower* by H.M.D. Parker, p. 160.
48. For the 1937 strike, see A. McKinlay, 'The 1937 Apprentices Strike: Challenge from an Unexpected Quarter', *Journal of Scottish Labour History Society*, 20 (1985), pp. 14–32. See also A. McKinlay, 'From Industrial Serf to Wage Labourer: The 1937 Apprentice Revolt in Britain', *International Review of Social History*, 31 (1986), pp. 1–18.
49. PRO, CAB 102/19, pp. 459–60.
50. PRO, LAB 10/138, Report of a Committee of Inquiry into Stoppages of Work by Apprentices in Scotland, Deputy Chief Industrial Commissioner Scotland to IRD, Ministry of Labour, 18 March 1941.
51. PRO, CAB 102/19, p. 460.
52. PRO, LAB 10/138, Letter from C. Gilbert to M.A. Bevin, Ministry of Labour, 12 May 1941.
53. PRO, LAB 10/138, Memorandum from DCIC to IRD Ministry of Labour, 18 March 1941.
54. PRO, CAB 102/19, p. 460.
55. *Shipbuilding and Shipping Record*, 15 May 1941, p. 458.
56. PRO, LAB 10/363, DCIC Scotland, Weekly Reports, 10 January 1942 to 19 December 1942.
57. PRO, CAB 102/379.
58. PRO, CAB 102/9, p. 467.
59. Inman, *Labour*, p. 394.
60. *Fairplay*, 13 August 1942, p. 210. This Order was termed the Essential Work (Shipbuilding and Repairing) Order No. 2, dated 24 July 1942.
61. PRO, CAB 102/379, Clyde District Consultative Committee Meetings, 17 September 1942 and 21 May 1943.
62. Inman, *Labour*, p. 276.
63. Inman, *Labour*, p. 276.
64. NMM, SRNA/1/3994, Allegations of Slackness and Inefficiency in

Shipbuilding and Shiprepairing, excerpt from *Daily Sketch*, 15 May 1942.

65. NMM, SRNA/1/3994, Memorandum from R.L.H. to William Watson, Secretary of the SEF, 15 May 1942.
66. NMM, SRNA/1/3994, Excerpt from the *Daily Mirror*, 14 February 1942.
67. PRO, LAB 8/731, Report of a Shop Stewards Committee of the Walker Naval yard on the inefficient use of skilled labour, 1 September 1941. Memorandum of 2 February 1942 from J.H. Stokes, Ministry of Labour, Northern Regional Office, on the suggestion of forming 'Flying Squads'. Memorandum of 7 April 1942 reporting on the outcome of a meeting of employers who rejected the idea on 1 April and Admiral W.G. Maxwell's acceptance of the employers' position.
68. PRO, LAB 8/731, Shipbuilding and Shiprepair, Review of Labour Requirements on the North West Coast, Meeting in Mr Morgan's Room, 2 June 1942.
69. Inman, *Labour*, also cited in J.F. Clarke, *Building Ships on the North-East Coast: Vol. 2, c. 1914–1980* (Whitley Bay, 1997), p. 332.
70. F. Ennis and I. Roberts, 'The Time of Their Lives? Female Workers in North East Shipbuilding, 1939–1945', in A. Potts (ed.), *Shipbuilders and Engineers: Essays in Labour History in the Shipbuilding and Engineering Industries in the North East of England* (Durham, 1987), pp. 42–3.
71. PRO, CAB 102/434, p. 131.
72. *Shipbuilding and Shipping Record*, 13 March 1941. This Order was made by the Minister of Labour and National Service under Regulation 58 A of the Defence (General) Regulations, 1939.
73. W. Ashworth, *Contracts and Finance* (1953) pp. 106–17. House of Commons Parliamentary Papers 1942–43, Vol. II, Report from the Committee of Public Accounts, paras 10–17.
74. Ashworth, *Contracts*, p. 107.
75. Parliamentary Papers 1942–43, Vol. II, PAC Minutes of Evidence, Navy Appropriation Account, 9 June 1943, para. 4066.
76. Ashworth, *Contracts*, p. 109.
77. Ashworth, *Contracts*, pp. 112–13.
78. M.M. Postan, *British War Production* (1952), p. 291.
79. Postan, *British War*, pp. 291–2.
80. Roskill, *The War*, Vol. III, Appendix 22. For the course of the Battle of the Atlantic, see E.J. Grove (ed.), *The Defeat of the Enemy Attack on Shipping 1939–45* (Aldershot, 1998), *passim*.
81. *Shipbuilding and Shipping Record*, 2 July 1942.
82. NMM, SRNA/7/SEF 1/40, Report of AGM of SEF, 13 October 1942. The industry figure for women employed in federated shipbuilding and repairing establishments at June 1942 was approximately 3,404, against a figure of 2,900 in 1939.
83. NMM, SRNA/7/SEF 1/40, Report of AGM of SEF, 13 October 1942.
84. NMM, SRNA, 7, SC/1/14, AGM of Shipbuilding Conference, 20 March 1942.
85. Clarke, *Building Ships*, p. 332.
86. Buxton, *Warship Building*, p. 9.

87. PRO, ADM 1/11892, Labour in Naval and Mercantile Shipyards (Barlow) July 1942, and BT 28/319, Report to the Machine Tool Controller on the Equipment of Shipyards and Marine Engineering Shops, (Bentham) September 1942.
88. UGBAC, Scotts Shipbuilding and Engineering Co. Ltd Papers, GD319/12/1/10, Admiralty Correspondence, Shipbuilding Inquiry Report, Letter, 27 July 1942.
89. UGBAC, Scotts Shipbuilding and Engineering Co. Ltd Papers, GD319/12/1/10, Preliminary Meeting held by the Warship Group before Meeting with Controller, 20 August 1942.
90. UGBAC, Scotts Shipbuilding and Engineering Co. Ltd Papers, GD319/12/1/10, Shipbuilders Version of Meeting with Controller.
91. These were Clyde and Belfast, chaired by Sir Stephen Pigott of John Brown; Tyne, chaired by Robin Rowell of Hawthorn Leslie; North West, chaired by Sir Charles Craven and South, chaired by Sir John Thornycroft.
92. UGBAC, Scotts Shipbuilding and Engineering Co. Ltd Papers, GD 319/12/1/10, Meeting of the Warship Group at Carlisle, 25 August 1942.
93. C. Barnett, *The Audit of War* (1996 edition) p. 119.
94. PRO, CAB 102/445, Report to the Machine Tool Controller, 30 September 1942.
95. PRO, CAB 102/445, Report to the Machine Tool Controller, 30 September 1942.
96. PRO, CAB 102/442, Merchant Shipbuilding and Repairs, C.M. Kohan unpublished narrative, pp. 34 and 53.
97. For an appreciation of the sheer scale of the American shipbuilding effort, see F.C. Lane, *Ships for Victory: Shipbuilding under the U.S. Maritime Commission in World War II* (Baltimore, 1951). See also H.G. Fassett, *The Shipbuilding Business in the United States of America*, 1 (New York, 1948).
98. PRO, CAB 102/442, p. 34.
99. Hornby, *Factories*, p. 42.
100. Hornby, *Factories*, pp. 48–9.
101. *Shipbuilding and Shipping Record*, 16 March 1944.
102. By May 1941, women had 'an important place' in the boatbuilding yards. *Shipbuilding and Shipping Record*, 8 May 1941, p. 434.
103. PRO, LAB 8/662, Efforts to Secure the More Efficient Use of Labour in Clydeside Area Shipbuilding Industry, Memorandum from J. McMillen, Shipyard Labour Supply Office, and from V. Holmes to Mr Jenkins, Reporting Officer Scotland, 24 May 1943.
104. NMM, SRNA, SEF 1/3173A, Ministry of Labour Quarterly Returns in the Shipbuilding and Shiprepairing Industry: Statement by Districts of Percentages of Women in the Industry. These ranged from 0.12 per cent in Belfast to 7.53 per cent on Tyneside, and to 12.42 per cent in the East of Scotland.
105. PRO, CAB 102/434, p. 86.
106. PRO, CAB 102/434, p. 86.
107. *Fairplay*, 7 January 1943, p. 89. By this stage over 200 Canadian naval craft had been delivered, with 700 more on the way, and Canada had surpassed

the shipbuilding effort of Great Britain. Moreover, since the start of the war, contracts had been placed with Canadian yards for 172 cargo vessels and 14,000 small craft, and in the first nine months of 1942 Canada produced 50 10,000–ton freighters.

108. *Build the Ships: The Official Story of the Shipyards in Wartime* (1947) p. 63. This publication is particularly disappointing on the wartime role of women in the shipbuilding industry.
109. *Shipbuilding and Shipping Record*, 6 December 1945, pp. 561–3.
110. Parliamentary Papers, 1943–44 Vol. II, PAC Minutes of Evidence, 25 May 1943, paras 2963–92.
111. Ayre, 'Merchant Shipbuilding' p. 8.
112. Ayre, 'Merchant Shipbuilding' p. 8.
113. Ayre, 'Merchant Shipbuilding' p. 19.
114. Hornby, *Factories*, pp. 52–3.
115. Hornby, *Factories*, p. 52.
116. Ayre, 'Merchant Shipbuilding' p. 4.
117. PRO, LAB 8/662, Efforts to Secure the More Efficient Use of Labour Supply in the Clydeside Area and in the Shipbuilding Industry, Report for August 1944 from Mackie to Reporting Officer Scotland, Jenkins.
118. Mortimer, *History*, pp. 1 and 17–18.
119. PRO LAB 8/662, Efforts to Secure the More Efficient Use of Labour Supply in the Clydeside Area and its Shipbuilding Industry, letter from James Cameron, Secretary of the North East Coast Shiprepairers Association enclosing copy letter from Messrs Smith's Dock Limited to Wm. Watson of the SEF.
120. NMM, SRNA 7, SEF 4/99, Federation Circulars, 1944.
121. *Shipbuilding and Shipping Record*, 30 November, 14 December 1944 and 7 June 1945.
122. NMM, SRNA 4, P11/1, Improved Shipbuilding Practice. Future of the Shipbuilding Industry after the War: Notes of a Joint Meeting of Office Bearers and Immediate past Presidents of the Shipbuilding Conference and of the Shipbuilding Employers Federation, Edinburgh, 29 August 1944, pp. 1–3.
123. NMM, SRNA 4, P11/1, pp. 5–6.
124. NMM, SRNA 4, P11/1, p. 28.
125. For a consideration of these points, see L. Johnman, 'The Labour Party and Industrial Policy, 1940–1945' in N. Tiratsoo (ed.), *The Attlee Years* (1991), pp. 29–53.
126. NMM, SRNA 4, P11/1, Improved Shipbuilding Practice, Meeting of 13 October 1944.
127. NMM, SRNA 4, P11/1, Committee on Improved Shipbuilding Practice: Interim Report from Sub-Committee on Methods of Shipbuilding Construction, March 1945.
128. NMM, SRNA 4, P11/1, Committee on Improved Shipbuilding Practice.
129. NMM, SRNA 4, P11/3, Letter from Denny to Ayre, 6 February 1945.
130. Inman, *Labour*, p. 121.
131. NMM, SRNA 7, SEF 4/101, Conclusions reached by Conference and Works

Board Meeting, 25 May 1945.

132. Inman, *Labour*, p. 123.
133. *Statistical Digest of the War* (1951), Table 113.
134. Buxton, *Warship Building*, p. 18. It should be noted, however, that it was not for want of financial resources. Admiralty expenditure for naval shipbuilding and other production under the headings Propelling and Auxiliary Machinery, Hull Building, Armour Plates, Mountings, Guns, Ammunition and Torpedoes and Mines revealed expenditure of £864.9 million between 1939 and 1944. See the Admiralty Appropriation Accounts, 1939–1944.
135. Roskill, *The War*, Vol. III, part II, pp. 308 and 415.

4 The Missed Opportunity

1. PRO, MT9/3648, 'Notes of an Informed Discussion Between Representatives of the Shipbuilding and Ship Repairing Industries, the Admiralty and the Ministry of War Transport, on Post-war Problems', 12 December 1944.
2. PRO, CAB 24/207, 'Shipbuilding Industry: Relationship with Government after the War', Memorandum by the First Lord of the Admiralty, April 1944.
3. PRO, ADM1/17037, 'Shipbuilding' Memorandum by the Minister of War Transport, May 1944.
4. PRO, MT9/3956, 'Notes of a Meeting Held at the Admiralty to Discuss the Government's Scheme for the Shipbuilding Industry', 16 November 1944.
5. PRO, MT9/3956, Letter from the First Lord of the Admiralty to the Minister of War Transport, November 1944.
6. A. Murray Stephen, 'Full Employment in British Shipyards', Presidential Address to the Institution of Engineers and Shipbuilders in Scotland, 3 October 1944, *Transactions of the Institution of Engineers and Shipbuilders in Scotland* (Glasgow, 1944).
7. PRO, MT9/3848, 'Post War Prospects of British Shipbuilding', Note by the Minister of War Transport and the First Lord of the Admiralty, September 1944, see Appendix 10.1.
8. PRO MT9/3956, 'Shipbuilding' Memorandum initialled JL, 12 November 1945, and MT73/2 'Final Report of the Shipbuilding Committee', March 1946.
9. PRO, CAB 124, 'Shipbuilding Industry', Memorandum initialled TMN, 16 January 1946.
10. For the establishment of the Shipbuilding Advisory Committee see PRO, BT 199/1, 'The Shipbuilding Advisory Committee', June 1946.
11. *Shipbuilder and Marine Engine Builder*, March 1946.
12. *Shipbuilder and Marine Engine Builder*, February 1946.
13. *Shipbuilder and Marine Engine Builder*, January and November 1946.
14. *Shipbuilder and Marine Engine Builder*, April 1946.
15. Discussion on 'The Application of Modern Management Methods to the Shipbuilding Industry', *Transactions of the North East Coast Institute of Engineers and Shipbuilders*, 63 (1946–47).
16. *Shipbuilder and Marine Engine Builder*, January and April 1946; PRO, LAB

8/1086, 'German Shipbuilding', Memorandum by C.B. Coxwell, October 1946 and PRO BT 199/1, Shipbuilding Advisory Committee, Fourth Meeting, October 1946.

17. A. Slaven, 'British Shipbuilders: Market Trends and Order Book Patterns Between the Wars', *Journal of Transport History*, 3 (1981–82), pp. 43–56.
18. Slaven, 'British Shipbuilders'.
19. B.W. Hogwood, *Government and Shipbuilding: The Politics of Industrial Change* (Farnborough, 1979), pp. 22–4.
20. Slaven, 'British Shipbuilders', pp. 43–4, and G. Henning and K. Trace, 'Britain and the Motorship: A Case of the Delayed Adoption of New Technology?', *Journal of Economic History*, 35 (1975).
21. Lord Kyslant quoted in Henning and Trace, 'Britain and the Motorship', pp. 382–3.
22. Rowell has already been quoted. The official organ of the British Maritime Industries took a very robust line stating that 'the future of steam was never brighter than it is today'. *Lloyd's List*, 4 December 1946.
23. *Fairplay*, January 1953 and *Shipbuilder and Marine Engine Builder*, March 1953.
24. D. McKeown, 'Welding—The Quiet Revolution' in A. Slaven and F. Walker (eds), *European Shipbuilding: One Hundred Years of Change* (1983), pp. 100–5.
25. C. Tipper, *The Brittle Fracture Story* (1962), *passim*.
26. NMM, Ship's Box 666, 'Intermediate Aircraft Carrier 1942', Note of Controller's meeting held at the Grand Pump Room Hotel, Bath, March 1942.
27. Tyne and Wear Archives, 1811/216/9/1–3, Joseph L. Thompson & Sons Ltd., 54th Annual General Meeting, 31 August 1948, Chairman's notes.
28. Discussion on 'The Application of Modern Management Methods to the Shipbuilding Industry', *Transactions of the North East Coast Institute of Engineers and Shipbuilders*, 63 (1946–47).
29. *Shipbuilder and Marine Engine Builder*, February 1947.
30. *Glasgow Herald Trade Review*, 1948.
31. *Shipbuilder and Marine Engine Builder*, February 1947 and PRO, CAB 124/522. 'Correspondence from the Shipowners' Associations on the Post-war Shipping Position', February 1948.
32. National Museum of Labour History, Manchester, Labour Party Archives, LPA RD/146, 'Shipbuilding' September 1948.
33. PRO, MT 73/9, 'Report of the Shipbuilding Costs Committee', July 1949 and letters from M.A. Custance, Ministry of Transport to N.A. Guttery, Admiralty, 2 December 1949 and memorandum by M.A. Custance, 19 July 1949. The full sorry tale of the history and workings of this Committee is in NMM SRNA4 S/20 and S/20/1–S/20/1/10, 'Shipbuilding Costs', papers relating to the Holroyd Pearce Committee, 1945–1948.
34. PRO, MT 73/178, 'The Future of the Shipbuilding and Ship Repairing Industries', Memorandum by the First Lord of the Admiralty and Minister of Transport, 1 July 1949.
35. *Glasgow Herald Trade Review*, 1949.
36. *Glasgow Herald Trade Review*, 1950.
37. J. Ramsay Gebbie, 'The Shipyards: An Unprecedented Order Book', *Lloyd's*

List Annual Review, 1951.
38. *Glasgow Herald Trade Review*, 1951 and 1952.
39. See for example, the Smith's Dock Chairman, T. Eustace Smith, quoted in *Shipbuilder and Marine Engine Builder*, January 1953.
40. PRO, MT 65/326, 'British Shipping and Shipbuilding', Joint Report to the Prime Minister by the First Lord of the Admiralty and the Minister of Transport, June 1953.
41. PRO, ADM 205/106, 'Present Weakness of the Marine Engineering Industry and its Consequences', Appendix 4, extract from a private letter by Lester Goldsmith, 15 May 1953.
42. *Shipbuilder and Marine Engine Builder*, November 1954.
43. For a full analysis of the Norwegian case, cf. L. Johnman and H. Murphy, 'The Norwegian Market for British Shipbuilding' *Scandinavian Journal of Economic History*, 46 (1998).
44. *Shipbuilder and Marine Engine Builder*, November 1954.
45. *Shipbuilder and Marine Engine Builder*, December 1958.
46. Southampton City Archives (hereafter SCA), D/VT/24/5/13, Shipbuilding Conference Statistics, Letter from J.M. McNeill to Viscount Hailsham, April 1957.
47. SCA, D/VT/24/6/3, Meeting between the Controller of the Navy and the Chairman of the Warship Group, 26 September 1958.
48. SCA, D/VT/24/6/3, Meeting of the Controller and Warship Group held in the Admiralty, 22 October 1958.
49. PRO, CAB 128/2 CM(45) 56th, 27 November 1945.
50. A. Gorst and L. Johnman, 'British Naval Procurement and Shipbuilding, 1945–1964' in D.J. Starkey and A.G. Jamieson (eds), *Exploiting the Sea: Aspects of Britain's Maritime Economy Since 1870* (Exeter, 1998), p. 119.
51. PRO, ADM 167/124, 'Director of Plans Future Building Committee: 1945, New Construction Programme', March 1945.
52. PRO, ADM 167/141 B.953, 'Naval Construction Programme 1952–1956', 4 January 1955.
53. PRO, ADM 205/106, 'Machinery for the New Construction Programme, Maritime Supremacy: The State of British Marine Engineering', Report by the Deputy Engineer-in-Chief of the Fleet, July 1955.
54. PRO, ADM 1/26455, 'Shipbuilding', Note of a Meeting held at the Treasury, 25 April 1956.
55. PRO, T225/1159, Letter from the First Lord of the Admiralty to the Minister of Defence, 10 November 1956.
56. PRO, ADM 1/26487, Letter from A. Belch of the Shipbuilding Conference to the Secretary of the Admiralty, May 1956.
57. PRO, T225/880, 'Admiralty: Short-Term Reflationary Measures to Alleviate Unemployment', 13 November 1958.
58. PRO, BT291/76, Cabinet: Economic Policy Committee Meeting, EA (59) 14th, 8 July 1959.
59. *The Shipping World*, 15 January 1947, and *Shipbuilder and Marine Engine Builder*, February 1957.
60. A. Shonfield, *British Economic Policy since the War* (Harmondsworth, 1958),

p. 42.

61. Companies House, London, Yarrow and Company, Company Accounts, 1945–1956.
62. *The Shipping World*, 15 January 1947.
63. Her Majesty's Stationery Office (HMSO), Report of the Committee of Inquiry into Shipping, Cmnd 4337, May 1970, p. 11.
64. PRO, BT291/77, 'Immediate Prospects of Shipbuilding and Ship Repairing Industries', Memorandum by the Ministry of Transport, 12 January 1960.
65. PRO, ADM 1/27498, Letter from W. Darracott, Admiralty, to A.W. Peck, Treasury, February 1960.
66. PRO, T225/1281, Letter from W. Darracott, Admiralty, to A.W. Peck, Treasury, February 1960.
67. PRO, T234/155, Working Party on the Transport of Oil from the Middle East, 'The Future of British Shipbuilding', Note by the Admiralty and the Ministry of Transport, 11 February 1957.
68. PRO, T234/155, Working Party on the Transport of Oil from the Middle East, Final Report, 12 June 1957.
69. PRO, T234/155.
70. PRO, T234/155.
71. PRO, T234/155, Working Party on the Transport of Oil from the Middle East, 'Shipbuilding in the United Kingdom: Current Appraisal', Note by the Admiralty, September 1958.
72. PRO, MT72/273, Oil Transport from the Middle East, Minutes of a Meeting held in Sir Roger Makins' room, 12 November 1958 and Note of a Meeting in the Minister's room, 19 January 1959.
73. PRO, T234/630, 'The Shipbuilding Industry in the United Kingdom', Report by Officials, June 1959.
74. PRO, T234/630, Working Party on Shipbuilding Prospects and Remedies, Memoranda, Minutes and Correspondence, June 1959.
75. PRO, T234/60, 'Prospects of the Shipbuilding Industry', July 1959.
76. PRO, T234/60, 'Full Scale Government Inquiry', Note by the Secretary, 15 September 1959.
77. PRO, T234/60, 'Aid for Shipbuilding', Memorandum by J. Brunner, 21 October 1959.
78. PRO, T234/60.
79. PRO, T234/60.
80. PRO, BT291/49, 'Resignation of Sir Graham Cunningham from the Chairmanship of the Shipbuilding Advisory Committee', 16 March 1960.
81. *The Times*, 24 March 1960, cf. also PRO, DSIR 17/800, 'The Research and Development Requirements of the Shipbuilding and Marine Engineering Industries', 15 March 1960.
82. Hogwood, *Government*, pp. 45–9.
83. PRO, PREM 11/3991, Minute by the Minister of Transport to the Prime Minister, 9 August 1959.
84. PRO, BT291/77, 'Immediate Prospects of Shipbuilding and Ship Repairing Industries', 12 January 1960, and 'The Future of Shipbuilding on the North East Coast', 14 March 1960.

85. PRO, BT291/104, Official Committee on Shipbuilding, 'Shipbuilding at the End of 1961', 14 December 1960.
86. PRO, BT291/77, 'Brief Notes on "Remedies" for the Shipbuilding Industry', Memorandum by A.B. Birnie, 16 February 1960.
87. PRO, BT291/54, Shipbuilding Advisory Committee: Sub-Committee on Prospects. 'The Shipbuilding and Ship Repairing Industries in Relation to Other Industries in Great Britain', Note by the Ministry of Labour, July 1960, and BT292/67, Standing Committee on Shipbuilding and Ship Repairing, 'Survey of Employment in November 1961', December 1961.
88. PRO, BT291/57, Shipbuilding Advisory Committee, Sub-Committee on Prospects, Minutes of Eighth Meeting, 5 January 1961.
89. J.A. Milne, 'Shipbuilding: Outlook Uncertain', *Lloyd's List Annual Review*, 1958.
90. Quoted in *Fairplay*, 15 September 1960.
91. *Shipbuilder and Marine Engine Builder*, December 1960.
92. *Shipbuilder and Marine Engine Builder*, December 1960.
93. *Fairplay*, 26 July 1962.
94. NMM, SRNA 4/50/1/14, 'State of the Industry and Reports on Foreign Countries, 1956–1963', Executive Meeting, 1962.
95. NMM, SRNA 4/50/1/14.
96. PRO, BT291/1, Note of a Meeting with the Shipbuilding Conference, 31 July 1962.
97. PRO, BT291/1.
98. PRO, BT 291/1, Economic Policy Committee, 'Surplus Shipbuilding Capacity', Memorandum by the Parliamentary Secretary (Shipping) to the Minister of Transport, November 1962.
99. PRO, BT 291/1.
100. PRO, BT291/1, Economic Policy Committee, Minutes of the 30th Meeting, 24 October 1962.
101. PRO, BT291/1, Economic Policy Committee, 'Surplus Shipbuilding Capacity', Memorandum by the Minister of Transport, November 1962.
102. PRO, BT291/1.
103. PRO, BT291/1.
104. PRO, PREM 11/4482, Prime Minister's Personal Minutes, July 1963.
105. PRO, BT 291/1, Economic Policy Committee, 'Surplus Shipbuilding Capacity', Memorandum by the Minister of Transport, November 1962.
106. PRO, BT291/142, Inter-Departmental Committee on Contraction of the Shipbuilding Industry, 'Facts about the Shipbuilding Industry', Note by the Ministry of Transport, 9 November 1962.
107. PRO, BT291/142, Note of the First Meeting, 13 November 1962.
108. Bank of England, EID 9/42, 'Shipbuilding in the United Kingdom', Memorandum by A.C. Darby, 1 March 1960.

5 Things Begin to Slide

1. PRO, Department of Scientific and Industrial Research, DSIR, 17/800, Research Council, Economic Committee, draft report, 15 March 1960, 'The

Research and Development Requirements of the Shipbuilding and Marine Engineering Industries'. See also *The Times*, 24 March 1960.

2. PRO, BT 291/54 and 57, Shipbuilding Advisory Committee, 'Sub-Committee on Prospects' 1961.
3. Ministry of Transport, 'Shipbuilding Orders Placed Abroad by British Shipowners'. Report to the Ministry of Transport by Messrs Peat, Marwick, Mitchell and Company, 1961.
4. PRO, CAB 27/190, Report of the Cabinet Committee on Unemployment and CAB 27/196 (CU) 651–700, memoranda, February to September, 1924.
5. See Chapter 2.
6. *Fairplay*, January 1951.
7. PRO, T236/2129, 'Norway, Review of December 1948 Agreement Covering Trade and Payments', telegram, Sir L. Collier, Oslo, to Foreign Office, London, 25 June 1949, and telegram from Foreign Office to Oslo, 28 June 1949.
8. Companies House, London; Company No. 493234, Accounts and Annual Reports of the Ship Mortgage Finance Company, Ltd, 1951–1969 (hereafter SMFC, A & AR).
9. For the full story of the Ship Mortgage Finance Company, see L. Johnman and H. Murphy, 'A Very British Institution! A Study in Under-capitalisation: The Role of the Ship Mortgage Finance Company in Post-delivery Credit Financing within Shipbuilding, 1951–1967', *Financial History Review*, 6 (1999), pp. 203–21.
10. P. Stokes, *Ship Finance: Credit Expansions and the Boom-Bust Cycle* (1992), pp. 13–14.
11. A. Shonfield, *British Economic Policy Since the War* (Harmondsworth, 1958), pp. 246–7.
12. PRO, MT 73/426, 'Committee on the Working of the Monetary System, Memorandum from the General Council of British Shipping', 11 April 1958.
13. *Financial Times*, 12 November 1953.
14. SMFC, A & AR, Chairman's Statement, 24 November 1953.
15. London School of Economics (LSE), Piercy Papers (PP), Industrial and Commercial Finance Corporation (ICFC) 9/196, 'The Shipbuilding Conference, AGM', 7 October 1953.
16. LSE, PP, ICFC 9/196, Memorandum from P.G. Wreford to the Chairman, 9 October 1953.
17. LSE, PP ICFC 9/196, Letter from the Shipbuilding Conference to the Economic Secretary to the Treasury, Reginald Maudling, 26 November 1953. For the efforts of the Shipbuilding Conference to garner such evidence, see NMM, SRNA 8/F1/9, 'Foreign Competition: Norway 1953–1977'.
18. PRO, MT 73/426, 'Committee on the Working of the Monetary System', 11 April 1958.
19. SMFC, A & AR, Chairman's Statement, AGM, 2 September 1960.
20. PRO, BT 291/2, 'The Shipbuilding Conference', Note of a Meeting with Vice-Admiral J. Hughes-Hallett, Joint Parliamentary Secretary to the Minister of Transport, 26 April 1963.
21. PRO, BT 199/13, Shipbuilding Advisory Committee, 'Prospects of the

Shipbuilding Industry', Note by the Ministry of Transport, 11 June 1963.

22. PRO, BT 199/13, ibid.
23. B. Hogwood, *Government and Shipbuilding: The Politics of Industrial Change* (Farnborough, 1979), p. 62.
24. Hogwood, *Government and Shipbuilding*, p. 62.
25. NMM, SRNA 8/S47, 'Norway', A/S Securitas Ltd, to Fairfield Shipbuilding and Engineering Co. Ltd, 11 May 1945.
26. NMM, SRNA 8/S47, 'Norway', J.C. Aird, Norwegian Shipping in 1947, 15 December 1947.
27. PRO, T 236/2129, Minutes, 53rd meeting of the Cabinet Overseas Negotiating Committee, ON(49), 28 June 1949.
28. For the expansion of the shipbuilding industries in Sweden and Japan, see K. Olsson, 'Big Business in Sweden: The Golden Age of the Great Swedish Shipyards, 1945–1974' *Scandinavian Economic History Review*, 53 (1995); T. Chida and P.N. Davies, *The Japanese Shipping and Shipbuilding Industries: A History of Their Modern Growth* (1990) and Y. Yamshita, 'Responding to the Global Market in Boom and Recession: Japanese Shipping and Shipbuilding Industries, 1945–1980' in S.P. Ville and D.M. Williams (eds), *Management, Finance and Industrial Relations in the Maritime Industries: Essays in International Maritime and Business History* (St John's, 1994). For a fuller analysis of the importance of the Norwegian market to British shipbuilding, see L. Johnman and H. Murphy, 'The Norwegian Market for British Shipbuilding, 1945–1967', *Scandinavian Economic History Review*, 46, 2 (1998).
29. NMM, SRNA 8/S47, 'Norway', Shipbuilding Capacity: Scandinavia, Mr Rae, 21 April 1959.
30. NMM, SRNA 8/S47, 'Norway', British Shipbuilding Exports: The Norwegian Market, Report by L.S. Holt, 28 September 1967.
31. NMM, SRNA 8/S47, 'Norway', Letter from R.B. Shepheard to Principals of Certain Member Firms, 24 July 1967.
32. NMM, SRNA 8/S47, 'Norway', British Shipbuilding Exports: The Norwegian Market Report by L.S. Holt, 28 September 1967.
33. NMM, SRNA 8/S47, 'Norway'.
34. NMM, SRNA 8/S47, 'Norway'.
35. NMM, SRNA 8/S47, 'Norway'.
36. These issues are fully explored in A. Slaven, 'British Shipbuilders: Market Trends and Order Book Patterns between the Wars', *Journal of Transport History*, 3 (1982), pp. 37–61; and A. Slaven, 'Marketing Opportunities and Marketing Practices: The Eclipse of British Shipbuilding, 1957–1976' in L.R. Fischer (ed.), *From Wheel House To Counting House: Essays in Maritime Business History in Honour of Professor Peter Neville Davies* (St John's, 1992) pp. 125–51.
37. Shipbuilding Inquiry Committee, 1965–1966, *Report*, Cmnd 2937 (1966) p. 51.
38. Slaven, 'Marketing Opportunities', p. 139.
39. J.R. Parkinson, *The Economics of Shipbuilding in the United Kingdom* (Cambridge, 1960), p. 61.
40. On these various aspects, see, NMM, SRNA 4/E40–E42/4, and NMM, SRNA 8/M41–M41/18. See also NMM, SRNA 8/E40–E41/2, and also a

huge amount of individual country studies scattered throughout SRNA 8.

41. Slaven, 'Marketing Opportunities', p. 142.
42. F.A.J. Hodges, 'The Application of Modern Management Methods to the Shipbuilding Industry', *Transactions of the North East Coast Institute of Engineers and Shipbuilders*, 63 (1946–47).
43. Discussion on 'The Application of Modern Management Methods', *Transactions of the North East Coast Institute of Engineers and Shipbuilders*, 63 (1946–47).
44. M. Moss and J.R. Hume, *Shipbuilders to the World: 125 Years of Harland and Wolff, Belfast 1861–1986* (Belfast, 1986), pp. 378–81.
45. *Fairplay*, 5 October 1961, and *Shipbuilder and Marine Engine Builder*, March 1963.
46. Quoted in Slaven, 'Marketing Opportunities', p. 143.
47. Committee of Inquiry into Shipping, *Report* (1970) p. 55.
48. A T.2 equivalent is defined as a vessel of 16,765 dwt with a service speed of 14.5 knots. A tanker's T.2 equivalent is thus:

$$\frac{\text{service speed in knots} \times \text{deadweight tonnage}}{14.5 \times 16{,}765}$$

49. Committee of Inquiry into Shipping, *Report* (1970) pp. 153–5.
50. Shipbuilding Inquiry Committee, *Report* (1966), Appendix D.

6 Death by Inquiry

1. There is a substantial literature on this period, but see in particular: P. Foot, *The Politics of Harold Wilson* (Harmondsworth, 1968), pp. 119–54; K. Middlemas, *Power, Competition and the State*, Vol. 2 (1990), pp. 57–111 and 150–86; and R. Coopey, S. Fielding and N. Tiratsoo (eds), *The Wilson Governments, 1964–1970* (1993) in particular the essays by Fielding, Horner, Woodward and Coopey, pp. 29–47, 48–71, 72–101 and 102–22 respectively. A. Cairncross, *The British Economy since 1945* (Oxford, 1992), pp. 151–81; F. Blackaby (ed.), *British Economic Policy 1960–1974* (Cambridge, 1978); and J. Tomlinson, *Public Policy and the Economy Since 1900* (Oxford, 1990), p. 253.
2. Quoted in Fielding in Coopey, Fielding and Tiratsoo (eds), *The Wilson Governments,* p. 40.
3. *Japanese Shipyards*, Board of Trade, 1965, A Report on the Visit of the Minister of State (Shipping) in January 1965.
4. Shipbuilding Inquiry Committee (SIC), 1965–1966, *Report*, Board of Trade, Cmnd 2937 (1966).
5. SIC, *Report*, pp. 162 and 164–5.
6. University of Cambridge Library (UCL), Vickers Archive, 505, Internal Memorandum, 12 July 1966; Memorandum on the Geddes Report as Affecting VSG Barrow-in-Furness, 3 May 1966.
7. SIC, *Report*, p. 88.
8. UCL, Vickers Archive, 505 'Geddes Report', Memorandum by D.F. Bruce and G.A. Smith, 10 May 1966.
9. NMM, SRNA7/SEF/2/27, Note on the Developments in Regard to Government's Participation in Management at Fairfield's, Memorandum by

N.A. Sloan, 10 December 1965.

10. PRO, BT291/120, 'Inter-Departmental Committee on the Fairfield Shipbuilding and Engineering Co. Ltd', Letter from the Chairman of the SIC, Reay Geddes, to the President of the Board of Trade, Douglas Jay, 31 December 1965; Letter from Geddes to Jay, 5 January 1966; and Letter from Jay to Geddes, 6 January 1966.
11. SIC, *Report*, pp. 45–50.
12. SIC, *Report*, pp. 88–92, 146–150 and 154–6.
13. R.H.S. Crossman, *The Diaries of a Cabinet Minister: Vol. 1, Minister of Housing, 1964–66* (1976), p. 582.
14. British Parliamentary Papers (BPP), 1967–68, Shipbuilding Industry Board, First Report and Accounts, 24 July 1968.
15. UGBAC, UCS1 9/6, 'Geddes Report: Meeting with Certain Clydeside Shipbuilders', Diary note by Lord Aberconway, 24 April 1966.
16. I. Johnston, *Ships for a Nation, 1847–1971: John Brown & Company, Clydebank* (West Dumbartonshire, 2000), pp. 262 and 272.
17. UGBAC, UCS 1 9/6, 'Geddes Report: Meeting with Certain Clydeside Shipbuilders', Diary note by Lord Aberconway, 16 April 1966.
18. UGBAC, UCS 1 9/6, 'Clyde Talks', Diary note by Lord Aberconway, 16 May 1966, and covering note, 18 May 1966.
19. UGBAC, UCS 1 9/6, 'Visit to Clydebank of Mr W. Swallow, Chairman, Shipbuilding Industry Board', 2 September 1966, Note by J. Rannie, 5 September 1966.
20. UGBAC, UCS 21/9, Minutes of a Board Meeting, Fairfield (Glasgow), 16 July 1967. Just how Anthony Hepper emerged in this position has long been a mystery. In 1999 Lewis Johnman conducted an interview with Sir Eric Yarrow, who responded to the question as to how Hepper had become the first Chairman of Upper Clyde Shipbuilders: 'you would probably have to ask Wedgwood Benn about that'. Sir Eric Yarrow interview, 15 April 1999. However, the following letter appears in the papers of the merchant bankers, Barings:

    If I had thought you were a runner, or even a starter, I would readily make such a nomination myself. I think some time ago I did make some suggestion . . . but I did not really think you could throw off the shackles of your SIB commitments and your longer term Tilling arrangement. If you do feel like accepting this challenging appointment . . . one major and very difficult hurdle would be overcome. I would like to assure you that such an appointment not only would have my full support, but with an acceptable financial formula might well influence our Board of Directors in any recommendations they might put forward to our shareholders on any merger proposals for our shipbuilding subsidiary company.

    Baring/ING Archives, UCS/201121 and 201122, Letter from Sir Eric Yarrow to A.E. Hepper, 18 May 1967.
21. UGBAC, UCS 5 2/18, 'Upper Clyde Working Party Report', July 1967.
22. PRO, FV37/5, Shipbuilding Industry Board (SIB) 'Tyne Grouping', Note of a Meeting, 21 December 1966.

23. PRO, FV37/5, 'Tyne Grouping', Note of a Meeting, 2 February 1967.
24. PRO, FV37/5, 'Tyne Grouping', Note of a Meeting, 22 February 1967.
25. PRO, FV37/5, 'Tyne Grouping', Letter from Sir Matthew Slattery to Sir William Swallow, 17 May 1967.
26. PRO, FV37/5, 'Tyne Grouping', Extract from *The Times*, 6 June 1967, reporting on the results of Sears Holdings Ltd, which owned the Furness Shipbuilding Company.
27. PRO, FV37/5, 'Tyne Grouping', Note of a Meeting, 8 June 1967.
28. PRO, FV37/5, 'Tyne Grouping', Note for File, 4 October 1967.
29. PRO, FV37/5, 'Tyne Grouping', Director's Visit to the Tyne, 8 January 1968.
30. PRO, FV 37/78, 'Tyne Grouping', Proposed acquisition of Furness Shipbuilding Co. Ltd by Swan Hunter and Tyne Shipbuilders Ltd, 20 September 1968.
31. PRO, FV37/5, 'Tyne Grouping', Note on a Visit to North-East Shipyards by C.H. Baylis, 19 June 1968.
32. PRO, FV 37/78, 'Tyne Grouping', Proposed acquisition of Furness Shipbuilding Co. Ltd by Swan Hunter and Tyne Shipbuilders Ltd, 20 September 1968.
33. PRO, FV37/2, 'Wear Grouping', Letter from Charles Longbottom, Austin and Pickersgill, to Sir William Swalllow, 3 February 1967.
34. PRO, FV37/71, 'Wear Grouping', Summary of Working Party Report, 6 October 1967.
35. PRO, FV37/76, 'Wear Grouping', Principal Comments by the Steering Committee of the Working Party Reports, 5 February 1968.
36. PRO, FV37/77, 'Wear Grouping', Progress in Grouping, 22 August 1968.
37. PRO, FV37/78, 'Wear Grouping', Letter from Sir H. Wilson Smith, Doxford's, to B. Barker, SIB, 10 September 1968.
38. PRO, FV37/38, 'Wear Grouping', Visit of Board Members to Wear Shipyards, 18/19 September 1968 and accompanying briefs for visits to Bartram's, Austin and Pickersgill and Doxford's.
39. SIC, *Report*, p. 164.
40. House of Commons, Official Report, Sixth Series, Vol. 733, Col. 1401.
41. UGBAC, Scott Lithgow Papers, GD323 1/5/5, Letter from Sir Eric Yarrow to Michael Sinclair Scott, 13 June 1966; Letter from Lord Aberconway to Scott, 16 June 1966, and Note of a meeting between Scott and Sir William Lithgow, 22 June 1966.
42. PRO, FV 37/70, SIB Minutes, Minutes of a Board Meeting, 22 May 1967. Tony Hepper, the SIB member responsible for the Working Party on Upper Clyde Shipbuilders, commented that three main points were likely to be involved in its proposals, one of which was the inclusion of the Lower Clyde firms in any viable merger.
43. UGBAC, GD 323 1/15/2, Annex to Report on Lower Clyde Shipbuilders, 10 February 1967. Scott of Bowling was purchased in May 1965 at a price of £63,500 and Greenock Dockyard in April 1966 at a purchase price of £264,000. Scott's undertook to complete the contracts on hand for the vendors by being reimbursed for the direct cost of labour and materials, plus a contribution to overheads. The offer was conditional on British and

Commonwealth placing an order for a mixed cargo vessel worth £1.52 million with Scott's which it duly did.

44. UGBAC, GD 323 1/15/2, Lithgows had bought the Glen yard of the William Hamilton yard from its co-partners, Brocklebanks, in 1960, and the remaining two-thirds of the shares in Ferguson Brothers in June 1963.
45. PRO, FV 37/70 SIB, Minutes of a Board Meeting, 9 October 1967.
46. PRO, FV 37/70 SIB, Letter from Benn to Sinclair Scott, 3 December 1968.
47. PRO, FV 37/70 SIB, Letter from Benn to Sinclair Scott, 3 December 1968.
48. SIB, Report and Accounts for the year ended 31 March 1970.
49. *Greenock Telegraph*, 14 January 1970.
50. UGBAC, GD 323 1/1/53, Letter from Sir Eric Yarrow to Michael Sinclair Scott, 9 February 1970.
51. UGBAC, GD 323 1/1/53, Letter from Ross Belch to Stephen Spain, Mintech, 19 February 1970.
52. UGBAC, GD 323 1/1/53, Scott Lithgow Limited, Summary of Negotiations for the Acquisition of Yarrow and Company Limited, 1 February 1971.
53. PRO, FV 37/137, SIB, Scott Lithgow, Approval of Draft Loan Agreement and Security Documents, 27 August 1970.
54. PRO, FV 37/137, SIB, Note of a Meeting with N. Ridley, 22 September 1970.
55. UGBAC, UCS2 1/9, Minutes of a Board Meeting, 7 August 1967.
56. UGBAC, UCS2 1/9, Minutes of a Board Meeting, 31 October 1967.
57. UGBAC, UCS2 1/9, Minutes of a Board Meeting, 6 February 1968.
58. S. Paulden and B. Hawkins, *Whatever Happened at Fairfield's?* (1969), pp. 181–90.
59. Paulden and Hawkins, *Whatever Happened?*, p. 185.
60. Paulden and Hawkins, *Whatever Happened?*, p. 192.
61. Shipbuilding Industry Board (SIB), First Report and Accounts, 24 July 1968.
62. UGBAC, UCS 5 2/3, Capital Investment Programme, Memorandum by J.R. Duff, 4 September 1968.
63. UGBAC, UCS 5 2/3, Letter from A.E. Hepper (UCS) to Sir William Swallow (SIB), 6 March 1969; and Letter from B. Barker (SIB) to A.E. Hepper (UCS), 7 March 1969.
64. UGBAC, UCS 5 2/3, Letter from A.E. Hepper (UCS) to Sir William Swallow (SIB), 10 March 1969.
65. UGBAC, UCS 5 2/3, Notes of a Meeting at Fitzpatrick House, 23 April 1969; and Memorandum 'Upper Clyde Shipbuilders', 24 April 1969.
66. UGBAC, UCS 5 2/3, Upper Clyde Shipbuilders, Summary of Corporate Plan, 2 May 1969.
67. T. Benn, *Office Without Power: Diaries 1968–1972* (1988), Entry for 17 March 1969, p. 155.
68. UGBAC, UCS 5 2/3, Statement by the UCS Board, 18 June 1969.
69. UGBAC, UCS 5 2/3, Statement by the Chairman, Mr A.E. Hepper, 1 August 1969.
70. A. Roth, *Heath and the Heathmen* (1972), pp. 211–12.
71. I. Gilmour and M. Garnett, *Whatever Happened to the Tories* (1998), p. 248.
72. J.R. Hume and M.S. Moss, *Shipbuilders to the World: 125 Years of Harland and Wolff, Belfast 1861–1986* (Belfast, 1986), pp. 400–45, and K. Warren, *Steel, Ships*

and Men (Liverpool, 1998), pp. 283–91.
73. J. McGill, *Crisis on the Clyde* (1973), p. 61.
74. McGill, *Crisis*, p. 71.
75. Roth, *Heath*, pp. 218–21, and Gilmour, *Whatever*, p. 258.
76. Sir Eric Yarrow, interview with Lewis Johnman, 15 April 1999.
77. McGill, *Crisis*, pp. 84–5.
78. Quoted in McGill, *Crisis*, pp. 87–8.
79. McGill, *Crisis*, pp. 93–4.
80. HMSO, *Report of the Advisory Group on Shipbuilding on the Upper Clyde*, HC544, 29 July 1971, pp. 1–3.
81. There is a huge literature on this subject. See, in particular, J. Foster and C. Woolfson, *The Politics of the UCS Work-In* (1986); W. Thompson and F. Hart, *The UCS Work-In* (1972); F. Herron, *Labour Market in Crisis: Redundancy at Upper Clyde Shipbuilders* (1975); A. Buchan, *The Right to Work* (1972); and McGill, *Crisis*.
82. Department of Trade and Industry, 'Shipbuilding on the Upper Clyde', Report of Hill Samuel & Co. Ltd, HMSO, Cmnd 4918 (1972) p. 115.
83. McGill, *Crisis*, pp. 127–8.
84. SIC, *Report*, p. 76.

7 Things Fall Apart

1. Shipbuilding Inquiry Committee, 1965–1966, *Report*, p. 43.
2. SIC, *Report*, p. 152.
3. SIC, *Report*, pp. 22–6, 27–32, 132–3 and 141–2.
4. PRO, BT 186/32, SIC, Replies to Questionnaire from the Ben Line, Cunard, British and Commonwealth Shipping and Shaw Saville and Albion, August–October 1965.
5. PRO, BT 186/32, SIC, 'Factors Affecting the Competitiveness of the British Shipbuilding Industry', Evidence of Captain H. White, Technical Director of J. and J. Denholm (Management) Ltd, Glasgow, 29 July 1965.
6. PRO, BT 186/32, SIC, Statement by Mr Alan Beaton, PA Management Consultants Ltd., 20 September 1965.
7. PRO, BT 186/32, SIC, Memorandum by Elliott Marine Automation Ltd, September 1965, and Letter from J.H. Carruthers and Co. Ltd, 6 October 1965.
8. PRO, BT 186/32, SIC, Note on a Meeting with Sir Stewart MacTier of Alfred Holt and Co., 6 July 1965, and Notes by Alfred Holt and Co., Liverpool, 11 August 1965.
9. PRO, BT 186/32, SIC, Peninsular and Oriental Steam Navigation Company, Evidence for SIC, August 1965.
10. PRO, BT 186/32, SIC, BP Tanker Company, Completed Questionnaire to SIC, undated but probably August–September 1965.
11. SIB, Report and Accounts 1971.
12. Booz-Allen and Hamilton, *British Shipbuilding 1972* (1973), p. 102.
13. SIC, *Report*, p. 29.
14. Booz-Allen and Hamilton, *British Shipbuilding*, p. 102.

15. Booz-Allen and Hamilton, *British Shipbuilding*, p. iii.
16. Booz-Allen and Hamilton, *British Shipbuilding*, pp. 75, 102, 116–20 and 139–58.
17. Booz-Allen and Hamilton, *British Shipbuilding*, pp. 59–67.
18. M. Davies, *Belief in the Sea* (1992), p. 233.
19. For the determined campaign against the nationalisation of the industry see NMM SRNA8/N5/2 to N5/2/19, 1973–76.
20. Davies, *Belief*, p. 240.
21. J.F. Clarke, *Building Ships on the North-East Coast* (Whitley Bay, 1999), pp. 443–54; M. Moss and J.R. Hume, *Shipbuilders to the World: 125 Years of Harland and Wolff, Belfast 1861–1986* (Belfast, 1986), pp. 446–62 and Davies, *Belief*, pp. 242–3.
22. For the impact of the international crisis on shipping, see G. Harlaftis, *A History of Greek-Owned Shipping* (1996), pp. 246–68; S. Tenold and H.W. Nordvik, 'Coping with the International Shipping Crisis of the 1970s: A Study of Management Responses in Norwegian Oil Tanker Companies', *International Journal of Maritime History*, 8, 2 (1996), pp. 33–69; M. Porter, 'The Oil Tanker Shipping Industry' in M. Porter (ed.), *Cases of Competitive Strategy* (New York, 1983); and also M. Ratcliffe, *Liquid Gold Ships: A History of the Tanker, 1859–1984* (Colchester, 1985). On the international economic crisis in general, see P. Armstrong, A. Glynn and J. Harrison *Capitalism Since 1945* (Oxford, 1991), pp. 210–61; A. Maddison, *Dynamic Forces in Capitalist Development* (Oxford, 1991), pp. 177–192; and H. Van Der Wee, *Prosperity and Upheaval: The World Economy 1945–1980* (Harmondsworth, 1987), pp. 479–512.
23. D. Todd, *Industrial Dislocation: The Case of Global Shipbuilding* (1991), pp. 3–10 and 32–86.
24. B. Strath, *The Politics of De-industrialisation: The Contraction of the West European Shipbuilding Industry* (1987), p. 22.
25. See especially 'The Directive on Shipbuilding' *Official Journal of the European Communities*, 27 July 1972; the Fourth Directive on Shipbuilding, 4 April 1978; the Council Resolution on the Reorganisation of Shipbuilding, 28 April 1981, and the Council Directive Extending the Fifth Directive, 1984.
26. BPP, Trade and Industry Committee, First Report, 'British Shipbuilders', Minutes of Evidence, 2 December 1981.
27. For the progress of the bill in parliament, see Davies, *Belief*, pp. 239–61.
28. Tyne and Wear Archives (TWA), Swan Hunter Papers, uncatalogued, Minutes of Meeting of Directors of Swan Hunter Shipbuilders Limited, 12 September 1975.
29. Official Report, CM1 (London) 2 December 1975.
30. Official Report, XXXIII, 29 November 1982.
31. TWA, Swan Hunter Papers, uncatalogued, Minutes of Meeting of Directors of Swan Hunter Shipbuilders Limited, 19 July 1976.
32. P.A. Milne, 'Shipbuilding Operations in British Shipbuilders Limited', *Transactions of the North East Coast Institute of Engineers and Shipbuilders*, 96 (1979–80).
33. Davies, *Belief*, p. 259.

34. BPP (1981–82), Trade and Industry Committee, 'British Shipbuilders', Memorandum submitted by British Shipbuilders, 2 December 1981.
35. See, for example, Scottish Development Department, *North Sea Oil: Production Platform Towers: Construction Sites* (Edinburgh, 1973) and International Management and Engineering Group (hereafter IMEG), *Study of Potential Benefits to British Industry from Offshore Oil and Gas Developments* (London, 1972). There is also a massive literature on the alleged failure to exploit properly North Sea Oil by various Governments. See, in particular, G. Arnold, *Britain's Oil* (1978), A. Jones, *Oil: The Missed Opportunity* (1981), A. Hamilton, *North Sea Impact: Offshore Oil and the British Economy* (1978), C. Harvie, *Fool's Gold: The Story of North Sea Oil* (Harmondsworth, 1995) and A. Jamieson, 'British OSV Companies in the North Sea, 1964–1997' *Maritime Policy and Management*, 25, 4 (1998), pp. 305–12.
36. For the initial attempts to interest the industry in the offshore market, see NMM, SRNA8 E36/21 Enquiries: Drilling Rigs, 1961–66 and unmarked file, USA and UK. For a full consideration of the shipbuilding industry's approach to the offshore market, see L. Johnman and H. Murphy, 'A Triumph of Failure: The British Shipbuilding Industry and the Offshore Structures Market, 1960–1990: A Case Study of Scott Lithgow Limited', *International Journal of Maritime History* (forthcoming, 2002).
37. IMEG, *Study of Potential Benefits*, pp. 9 and 85–96.
38. *Greenock Telegraph*, 3 June 1981.
39. BPP (1981–82) Trade and Industry Committee, Minutes of Evidence, 2 December 1981.
40. Official Report, XLVI, 28 July 1983. Scott Lithgow had lost £155 million between 1977 and 1983. Calculated from BPP (1983–84), Scottish Affairs Committee, Second Report, 'Scott Lithgow Limited: The Economic and Social Consequences of Closure', Memorandum submitted by BS.
41. By definition a state corporation could not go bankrupt but we feel sure that the reader will take the point.
42. G. Maynard, *The Economy Under Mrs Thatcher* (Oxford, 1988) p. 28.
43. Official Report, Vol. LI, 19 December 1983.
44. British Shipbuilders, Report and Accounts, 1983–84, Chairman's Statement.
45. British Shipbuilders, Report and Accounts, 1983–84, Chairman's Statement.
46. Official Report, LIII 1 February 1984 and LVII, 3 April 1984.
47. BPP (1984–85), Trade and Industry Committee, 'British Shipbuilders', Minutes of Evidence, 13 February 1988. As well as the Scott Lithgow privatisation, Day sold the Ship-Repairing Division to a combination of management buy-outs and takeovers.

8 Privatised Unto Death

1. There is a huge literature on this topic, but see, in particular, C. Johnson, *The Economy Under Mrs Thatcher 1979–1990* (Harmondsworth, 1991); F. Green (ed.), *The Restructuring of the UK Economy* (1989); and J. Michie (ed.), *The Economic Legacy 1979–1992* (1992), and the bibliographies therein for privatisation in particular and the period generally.

2. Quoted in Johnson, *The Economy*, p. 179.
3. Johnson, *The Economy*, p. 156.
4. BBC Television, *Close Up North*, 'The Secret Deal That Sank a Shipyard', 1 September 1993; Sir Robert Atkinson, *The Development and Decline of British Shipbuilding: Some Thoughts and Comments* (1999,) p. 29; and Sir Robert Atkinson, interview with Lewis Johnman, 17 December 1999.
5. BPP (1984–85), Trade and Industry Committee, 'British Shipbuilders', Minutes of Evidence, 13 February 1985.
6. Official Report, LXXXII, 5 July 1985.
7. For the re-statement of policy, see Official Report, LXVII, 23 July 1973.
8. BPP (1984–85), Trade and Industry Committee, 'British Shipbuilders', Minutes of Evidence, 13 February 1985.
9. BPP (1984–85), Committee of Public Accounts, Thirty Fifth Report, 'Design and Procurement of Warships', 22 July 1985.
10. BPP (1984–85), Trade and Industry Committee, Trade and Industry Committee, 'British Shipbuilders', Minutes of Evidence, 13 February 1985.
11. BPP (1984–85), Committee of Public Accounts, Design and Procurement, 22 July 1985.
12. Official Report, XLI, 19 April 1983.
13. BPP (1984–85), Committee of Public Accounts, Design and Procurement, 22 July 1985.
14. Official Report, XCV, 25 March 1985.
15. BPP (1988–89), Trade and Industry Committee, Third Report, 'The Privatisation of Harland and Wolff plc', and Official Report, XCVI, 24 March 1986.
16. Official Report, XCVI, 24 April 1986.
17. Quoted in E.J. Grove, *Vanguard to Trident: British Naval Policy Since World War Two* (Annapolis, 1989), pp. 345–6.
18. K. Speed, *Sea Change* (Bath, 1982) pp. 99–116 for his account of this. See also Grove, *Vanguard*, pp. 342–99.
19. Official Report, XCVIII, 15 July 1986.
20. Official Report, CXXXVI, 11 July 1988.
21. Official Report, CLXII, 19 December 1989.
22. Official Report, CLXXXII, 20 November 1990.
23. BPP (1990–91), Defence Committee, Third Report, 'Options for Change: Royal Navy', 27 February 1991.
24. Official Report, CCXI, 16 July 1992.
25. Official Report, CCXXIII, 11 May 1993.
26. BPP (1993–94), Committee of Public Accounts, Thirtieth Report, 'Ministry of Defence: The Award of the Landing Platform for Helicopters', 27 June 1994, Appendix 3, Minutes of Meetings between the Senior Management and the CSEU of VSEL, 16 March and 6 April 1993.
27. BPP (1993–94), Committee of Public Accounts, Thirtieth Report, 'Ministry of Defence: The Award of the Landing Platform for Helicopters', 27 June 1994, Appendix 3.
28. *Guardian*; and *Independent*, 30 August 1994.
29. Lord Weinstock, GEC, to Eddie Darke, APEX, Swan Hunter, 14 September

1994, copy of letter in possession of Lewis Johnman.

30. Monopolies and Mergers Commission, 'British Aerospace Public Limited Company and VSEL plc', A Report on the proposed merger, Cmd 2851 (1995), p. 7.
31. Monopolies and Mergers Commission, 'British Aerospace Public Limited Company and VSEL plc', A Report on the proposed merger, Cmd 2851 (1995), p. 7.
32. Official Report, LXIV, 25 July 1984.
33. BPP (1984–85), Trade and Industry Committee, 'British Shipbuilders', Minutes of Evidence, 24 July 1985.
34. BPP (1985–86), Committee of Public Accounts, Forty Third Report, 'Assistance to British Shipbuilders', 7 July 1986.
35. BPP (1985–86), Committee of Public Accounts, Forty Third Report, 'Assistance to British Shipbuilders', Appendix 1, 7 July 1986.
36. BPP (1985–86), Committee of Public Accounts, Forty Third Report, 'Assistance to British Shipbuilders', Examination of Witnesses, 21 May 1986.
37. Official Report, XCVII, 14 May 1985.
38. BPP (1985–86), Trade and Industry Committee, 'British Shipbuilders', Minutes of Evidence, 24 July 1985.
39. BPP (1985–86), Trade and Industry Committee, 'British Shipbuilders', Minutes of Evidence, 18 June 1986.
40. P. Stokes, *Ship Finance: Credit Expansion and the Boom–Bust Cycle* (1992), pp. 81–4.
41. Official Report, CII, 24 July 1986; and CVII, 8 December 1986.
42. BPP (1987–88), Trade and Industry Committee, 'British Shipbuilders', Minutes of Evidence, 14 June 1988.
43. BPP (1987–88), Trade and Industry Committee, 'British Shipbuilders', Minutes of Evidence, 14 June 1988.
44. Official Report, CLXXXVI, 27 June 1988. On the issue of SIF money, Clarke was asked if Govan would have qualified had it remained in the public sector; he replied 'probably not'. BPP (1988–89), Trade and Industry Committee, 'British Shipbuilders', 28 June 1988.
45. Official Report, CXLIII, 14 December 1988; and BPP (1988–89), Trade and Industry Committee, 'British Shipbuilders', Minutes of Evidence, 7 June 1989, 'Supplementary Memorandum submitted by the DTI'.
46. Official Report, CXLII, 7 December 1988.
47. Official Report, CXXXVII, 21 July 1988.
48. BPP (1988–89), Trade and Industry Committee, 'British Shipbuilders', Minutes of Evidence, 18 October 1989; and appendices, Sir Leon Brittain to the Chairman, 15 December 1989.
49. BPP (1988–89), Trade and Industry Committee, 'British Shipbuilders', Minutes of Evidence, Brittain to Bob Clay MP, 6 July 1989.
50. BPP (1988–89), Trade and Industry Committee, 'British Shipbuilders', Minutes of Evidence, Brittain to Tony Newton, Chancellor of the Duchy of Lancaster, 12 July 1989.
51. BPP (1988–89), Trade and Industry Committee, 'British Shipbuilders', Minutes of Evidence, 'Linkage Between NESL and Other BS Disposals', 18

October 1989.

52. BPP (1988–89), Trade and Industry Committee, 'British Shipbuilders', Minutes of Evidence, 14 June 1988.
53. BPP (1988–89), Trade and Industry Committee, 'The Privatisation of Harland and Wolff plc', 15 February 1989.
54. BPP (1988–89), Trade and Industry Committee, 'The Privatisation of Harland and Wolff plc', 15 February 1989.
55. BPP (1988–89), Trade and Industry Committee, 'The Privatisation of Harland and Wolff plc', 15 February 1989.
56. BPP (1988–89), Trade and Industry Committee, 'The Privatisation of Harland and Wolff plc', 15 February 1989.
57. Monopolies and Mergers Committee, 'British Aerospace Public Limited Company and VSEL plc, A Report on the proposed merger', Cmd 2851, p. 58.
58. Monopolies and Mergers Committee, 'British Aerospace Public Limited Company and VSEL plc, A Report on the proposed merger', Cmd 2851, p. 115.
59. Derived from Official Reports, various dates.
60. British Shipbuilders, *Annual Reports and Accounts*, 1977–78 to 1987–88; and BPP (1988–89), Trade and Industry Committee, 'British Shipbuilders', Minutes of Evidence, 14 June 1988.
61. North Tyneside Trades Council, *Shipbuilding—The Cost of Redundancy* (Newcastle-upon-Tyne, 1979), *passim*, and BPP (1985–86) Trade and Industry Committee, 'British Shipbuilders', Minutes of Evidence, 18 June 1986.
62. BPP (1985–86) Trade and Industry Committee, 'British Shipbuilders', Minutes of Evidence, 18 June 1986.
63. Official Report, CCXI, 11 July 1992.
64. PRO, ADM 116/5555, British Admiralty Mission to Washington, letter from Seal to Lillicrap, 21 March 1945.
65. PRO, ADM 116/5555, British Admiralty Mission to Washington, letter from Lillicrap to Seal, 23 March 1945.

Epilogue

1. Strategic Defence Review (Cm 3999), 1998.
2. *Lloyd's List*, 24 November 1999.
3. Scottish Office, *News Release*, 18 January 2000.
4. *The Observer*, 15 April 2001.
5. *The Guardian*, 11 and 12 July 2001.
6. Scottish Executive, *Press Release*, 18 July 2001.
7. *The Guardian*, 16 October 2001.

Bibliography

Primary Sources

Manuscripts

Bank of England:
Bankers' Industrial Development Trust
Securities Management Trust
Fairfield Shipbuilding and Engineering Company Ltd
Industry: Engineering and Shipbuilding, 1931–1961
Monetary Policy: Advances for Shipbuilding, 1959–1963
Chief Cashier's Private File
National Shipbuilders' Security Ltd

Bank of Scotland:
Fairfield's, 1962–1965

Churchill College, Cambridge:
Grigg Papers
Weir Papers

City of Glasgow Archives, Mitchell Library, Glasgow:
Barclay Curle
Clyde Shipbuilders' Association
Fairfield Shipbuilding and Engineering Company Ltd

Companies House:
Yarrow and Company
Ship Mortgage Finance Company

ING/Barings:
Upper Clyde Shipbuilders

London School of Economics and Political Science:
Piercy Papers

National Archives of Scotland:
Burntisland Shipbuilding Company
Fairfield Shipbuilding and Engineering Company Ltd

National Maritime Museum, Greenwich:
Shipbuilders' and Repairers' National Association
Shipbuilding Employers' Federation
Shipbuilding Conference
Director of Naval Construction Department

National Museum of Labour History, Manchester:
Labour Party Archives, Research Department

Public Record Office, Kew, London:
Admiralty:
Admiralty and Secretariat
Admiralty and Secretariat Case Books
War of 1939–45, War History Cases
Director of Naval Construction Reports
First Sea Lord Papers

Board of Trade:
Establishment Department: Registers and Indexes of Correspondence
Finance Department: Correspondence and Papers
Departmental Committees
Chief Industrial Adviser's Department
Ministry of Shipping: Correspondence and Papers
General Shipping Policy Files
Industries and Manufactures Department
Committee of Inquiry into the Shipbuilding Industry (Geddes Committee)
Shipbuilding Advisory Committee
Admiralty and Successors: Shipbuilding and Repair: Registered Files (SBR) Series

Cabinet:
Committee of Imperial Defence: Principal Supply Officers Committee
Official War Histories (Second World War) Civil Historical Section:
Registered Files
Minutes of Cabinet Meetings
Memoranda
Home Affairs Committee
Cabinet Committees
Economic Advisory Council

Ministry of Labour:
Correspondence and Papers
Employment
Industrial Relations

Department of Scientific and Industrial Research:
Registered Files: General Series

Ministry of Transport:
Marine: Correspondence and Papers
Shipping Policy

Prime Minister's Office:
Correspondence and Papers

Ramsay MacDonald:
Papers

Ministry of Reconstruction:
Records

Ministry of Technology:
Shipbuilding Industry Board: Registered Files (SIB) Series

Treasury:
Finance Files
Supply Files
Chancellor of the Exchequer's Office, Miscellaneous Papers
Hopkins Papers
Trades Facilities Acts Advisory Committee
Home and Overseas Planning Staff Division

Southampton City Records Office, Southampton:
John I. Thornycroft

Tyne and Wear Archives Service, Newcastle-upon-Tyne:
Joseph L. Thompson and Sons Ltd
Swan Hunter Papers

University of Cambridge Archives:
Vickers Papers

University of Glasgow, Business Archives Centre, Glasgow:
William Denny and Brothers
A. Stephen and Sons
William Beardmore and Company
William Simons and Company
Blythswood Shipbuilding Company Ltd
Barclay Curle
James Lamont and Company Ltd
UCS Work-in BBC Transcripts
Lithgows Ltd
Harland and Wolff
Yarrow & Co Ltd
Upper Clyde Shipbuilders

Scotts Shipbuilding and Engineering Co
Ferguson Brothers
Scott and Sons
Ailsa Shipbuilding Company Ltd
John Hastie and Sons Ltd
Sir James Lithgow Papers
Scott Lithgow Papers

Serial Publications:
Board of Trade Journal
British Shipbuilders Annual Reports and Accounts
Conway's All the World's Fighting Ships
Fairplay
Financial Times
Glasgow Herald Annual Supplement
Glasgow Herald Trade Review
Greenock Telegraph
Jane's Defence Weekly
Journal of Commerce, Shipbuilding and Engineering Edition
Keesings Contemporary Archives
Lloyd's List
Lloyd's List Annual Register
Lloyd's Register of Shipping
Ministry of Labour Gazette
Shipbuilder and Marine Engine Builder
Shipbuilding and Shipping Record
Shipping World
Smith's Dock Monthly
Stock Exchange Year Book
The Economist
The Guardian
The Motor Ship
The Naval Architect
The Shipping World
The Times
Transactions of the Institute of Shipbuilders and Engineers in Scotland
Transactions of the North East Coast Institute of Engineers and Shipbuilders
Transactions of the Royal Institute of Naval Architects
The Yorkshire Post Trade Review

Parliamentary Reports and Papers (in chronological order)

Engineering Trades After the War, Report of the Reconstruction Committee, 1918, Cmd 9073 XIII.
Shipping and Shipbuilding Industries after the War, Report of Board of Trade Committee, 1918, Cmd 9092 XIII.
Dockyards (Building Merchant Ships), Report of Admiralty Committee, 1920, Cmd 581 XXI.

National Expenditure, Reports of the Geddes Committee, 1922, Cmd 1581 LX I; 1922 Cmd 1582 IX; 1922 Cmd 1589 IX.

Industry and Trade, Final Report of the Balfour Committee, 1928–1929, 1929, Cmd 3282 VII.

Measures Proposed by the Government to Secure Reductions in National Expenditure, 1930–1931, 1931, Cmd 3952 XVIII.

National Expenditure. Report of the May Committee, 1930–1931, 1931, Cmd 3920 XVI I.

Unemployment Grants Committee, Reports under the Development (Loan Guarantees and Grants) Act 1929, Part II 1930–1931, Cmd 3744 XVII; 1931–1932, 1932, Cmd 4029 XIII; Final Report 1920–1932: 1932–1933, 1933, Cmd 4354 XV.

Merchant Shipping: Outline of a Scrapping and Rebuilding or Modernising Scheme for British Cargo Vessels Discussed Between the Board of Trade and Shipowners, 1933–1934, 1934, Cmd 4647 XVIII.

British Shipping (Assistance) Act 1935, Interim Report of Tramp Shipping Administrative Committee, 1934–1935, 1935, Cmd 5004 XI.

Tramp Shipping: Distribution of Subsidy under British Shipping (Assistance) Acts 1935 and 1936, 1936, Cmd 5129 XIII; 1936–1937, 1937, Cmd 5420 XVI.

Memorandum on Financial Resolution re. Continuance of Tramp Shipping Subsidy (1937), 1936–1937, 1937, Cmd 5357 XXI.

Fourth Report of the Tramp Shipping Advisory Committee, 1936–1937, 1937, Cmd 5363 XIII.

Agreement (15 February 1937) between Cunard White Star and the Treasury, 1936–1937, Cmd 5397 XXI.

Report of the Ships Replacement Committee of the Board of Trade, 1936–1937, 1937, Cmd 5459 XIII.

Fifth Report of the Tramp Shipping Advisory Committee, 1936–1937, 1938, Cmd 5555 XIII.

Agreement (16 February 1938) between Cunard White Star and the Treasury, 1937–1938, 1938, Cmd 5725 XXI.

Sixth Report of the Tramp Shipping Advisory Committee, 1937–1938, 1939, Cmd 5750 XIII.

Proposals for the Assistance of British Shipping, 1938–1939, 1939, Cmd 6060 XVIII.

Memorandum on the War-Time Financial Arrangements between the Government and British Shipowners, 1939–1940, 1940, Cmd 6218 VII.

Memorandum on Scheme for Purchase by British Shipowners of New Vessels Built on Government Account, 1941–1942, 1942, Cmd 6357 V.

Memorandum on Scheme for Purchase of Merchant Vessels by Allied Governments from His Majesty's Government, 1941–1942, 1942, Cmd 6373 V.

British Government Memorandum to OEEC re. British Shipping Policy, 1948–1949, 1949, Cmd 7572 XXXIV.

Agreement between the Government and Cunard re. Finance for Replacement of *Queen Mary*, 1960–1961, 1961, Cmd 1319 XVII.

Report of Shipbuilding Inquiry Committee (Geddes Committee), 1965–1966,

1966, Cmd 1937 VII.
Report of the Committee of Inquiry into Shipping (Rochdale Committee), 1969–1970, 1970, Cmd 4337 XXVII.
Shipbuilding on the Upper Clyde, Report by Hill Samuel and Company Ltd, 1971–1972, 1972, Cmd 4918 XXXVII.
White Paper on Industrial and Regional Development, 1971–1972, 1972, Cmd 4942 XVIII.
Treasury Minute on Third Report of Committee of Public Accounts, 1974–1975, 1975, Cmd 6298 XXII.
Treasury Minute on the First Report of Committee of Public Accounts, 1979–1980, 1980, Cmd 7788 XVI.
Defence White Paper, 'The Falklands Campaign: The Lessons', 1982, Cmd 8758.
Defence Estimates 1984, 1983–1984, Cmd 9227–1, 9227–11.
Defence Estimates 1985, 1984–1985, Cmd 9430–1, 9430–11.
Defence Estimates 1986, 1985–1986, Cmd 9763–1, 9763–11.
Government Response to the Fourth Report of the Defence Committee, 1986–1987, 1987, Cmd 228.
Defence Estimates 1987, 1986–1987, Cmd 101–1, 101–11.
White Paper, Merchant Shipping Bill, 1987–1988, Cmd 239.
Defence Estimates 1988, 1987–1988, Cmd 344–1, 344–11.
Defence Estimates 1989, 1988–1989, Cmd 671–1, 671–11.

House of Commons Sessional Papers

Trades Facilities Acts, Guarantees which Treasury have stated their willingness to give to 31.3.1922, 1922, (62) XVIII, published quarterly until 31.12.1926.
Trades Facilities Acts, Guarantees up to the completion of the scheme on 31.12.1927 and Account to 31.3.1927 of total sums issued under the Acts, and of the sums paid toward repayment, etc., 1927, (61) XIII, Annual accounts until 1949.
Contract between Postmaster General and Cunard and Oceanic Steam Navigation re. Sea Mails to the United States, 1928, (96) XIII.
Mail Contract (1 April 1936) between the Postmaster General and Cunard White Star, 1936, (92) XV.
Account of Advance under Pt II of the British Shipping (Assistance) Act 1935 for Period Ended 31 March 1937 and 31 March 1938, 1937–1938 (1) and (145) XVI.
Final Report of Treasury Guarantees Given Under the Trade Facilities Acts 1921–1926, 1962, (183) XXV.
Industrial Development Act 1966, Annual Report 1967–1968, 1968, (420) XXIV.
Shipbuilding Industry Board, First Report and Accounts for the Period Ended 31 March 1968, 1968, (361) XXX1.
Industrial Development Act 1966, Annual Report 1968–1969, 1969, (329) XXXIII.
Shipbuilding Industry Board, Report and Accounts for the Period Ended 31 March 1969, 1969, (326) XIV.
Industrial Reorganisation Corporation, Report for the Year Ended 31 March 1970, 1970, (310) XVI.
Shipbuilding Industry Board, Report and Accounts for the Year Ended 31 March

1970, 1970, (84) XIIX.
Industrial Development Act 1966, Annual Reports 1969–1970 and 1970–1971, 1971, (124 and 580) XXV.
Shipbuilding Industry Board, Report and Accounts for the Year Ended 31 March 1971, 1971 (554) XIIX.
Report of Advisory Group on Shipbuilding on the Upper Clyde, 1971, (544) XIIX.
Industrial Reorganisation Corporation, Final Report for the Period 1 April to 30 April 1971, 1971, (443) XXV.
Shipbuilding Industry Board, Report and Accounts for the Period 1 April to 31 December 1971, 1972, (316) XXXVII.
Committee of Public Accounts, Third Report Session 1971–1972, 1972, (447) XXIV.
Expenditure Committee on Public Money in the Private Sector, 1972 (347) XXVII.
Accounts re. Issues from the National Loans Funds 1971–1972, Statement of Guarantees under the Shipbuilding Industry Acts 1967 and 1971 for the Year Ended 31 March 1972, 1972, (155) XVI.
Industry Act 1972, Report for the Year Ended 31 March 1973, 1973, (429) XII.
Industry Act 1972, Report for the Year Ended 31 March 1974, 1974, (339) VII.
Committee of Public Accounts, Third Report Session 1974–1975, 1975, (374) XXII.
Court Line, Parliamentary Commissioner for Administration. Fifth Report Session 1974–1975, 1975, (498) XXIX.
Industry Act 1972, Report for the Year Ended 31 March 1975, 1975, (620) XIV.
Industry Act 1972, Report for the Year Ended 31 March 1976, 1976, (619) XVII.
Industry Act 1972, Report for the Year Ended 31 March, 1977, 1977 (545) XVI.
Northern Ireland Appropriation Accounts 1976–1977, 1977, 1977, (89) XXVI.
Industry Act 1972, Report for the Year Ended 31 March, 1978, 1978, (653) XVIII.
Report by the Auditor General on the Administration of the Intervention Fund, Appropriation Accounts, Vol. II, 1978, 1977–1978, (138) XI.
Northern Ireland Appropriation Accounts 1977–1978, 1978, (79) XI.
Industry Act 1972, Report for the Year Ended 31 March, 1979, 1979 (206) XVI.
Committee of Public Accounts, First Report Session 1979–1980, 1980, (173) XV.
Northern Ireland Appropriation Accounts 1978–1979, 1979, (314) XX.
Industry Act 1972, Report for the Year Ended 31 March, 1980, 1980, (772).
Committee of Public Accounts, Nineteenth Report Session 1979–1980, 1980, (737).
Committee of Public Accounts, Twenty Fourth Report Session 1979–1980, 1980, (763).
Industry Act 1972, Report for the Year Ended 31 March 1981, 1982, (460).
Accounts Relating to Issues from the National Loans Fund, 1981, (200).
Industry Act 1972, Report for the Year Ended 31 March 1982, 1982, (503).
Industry and Trade Committee, First Report on British Shipbuilders, (192).
Government Reply to the First Report of the Industry and Trade Committee, (381).
Industry Act 1972, Report for the Year Ended 31 March 1983, 1983, (72).
Defence Committee, First Report, 1983–1984, (436).
Defence Committee, Second Report, 1985–1986, (339).
Transport Committee, Interim Report on the Decline in the UK Registered Fleet,

(94–1).
Defence Committee, Fourth Report on Implementing the Lessons of the Falklands Campaign, (3451–11).
Trade and Industry Committee, 1987–1988, Evidence re. British Shipbuilders, (5191–11).
Transport Committee, Report on the Decline in the UK Registered Fleet, 1987–1988, (3031–11).
Government Observations on the Transport Committee's Report on the Decline of the UK Registered Fleet, 1987–1988, (681).
Defence Committee, Fourth Report on the Defence Requirements for Merchant Shipping, 1987–1988, (476).
Government Reply to the Fourth Report of the Defence Committee, 1987–1988, (674).
Trade and Industry Committee, Report on Privatisation of Harland and Wolff, 1988–1989, (1311–111).
Government Response to the Report of the Trade and Industry Committee on the Privatisation of Harland and Wolff and Short Brothers, 1988–1989, (353).

Government (Non-Parliamentary) and other Official Publications

Balfour Committee on Industry and Trade, Survey of the Metal Industries, 1927.
Balfour Committee on Industry and Trade, Survey of Factors on Industrial and Commercial Efficiency, 1927.
Report of the Department of Scientific and Industrial Research on the Research and Development Requirements of the Shipbuilding Industry, 1960.
Report of the Sub-Committee of the Shipbuilding Advisory Committee on Prospects of the Shipbuilding and Ship Repairing Industry, 1961.
Report of Peat, Marwick, Mitchell and Company, on Shipbuilding Orders Placed Abroad by British Owners, 1961.
International Management and Engineering Group, Study of Potential Benefits to British Industry from Offshore Oil and Gas Developments, 1972.
Scottish Development Department, North Sea Oil: Production Platform Towers: Construction Sites, 1973.
British Shipbuilding 1972, Report to the Department of Trade and Industry by Booz-Allen and Hamilton, 1973.
UK Ship Repairing Industry, Report to the Department of Trade by P.A. Management Consultants Ltd, 1974.
Interim and Final Reports of the Board of Trade Inspectors into the Affairs of Court Line Ltd, 1978.
British Shipbuilders: Annual Reports and Accounts 1977–1988.
British Shipping: Challenges and Opportunities.

European Economic Community

Council Aid to Shipbuilding Directive No. 72/273/EEC OJ No. L169/28 (1972).
Council Aid to Shipbuilding Directive No. 75/432/EEC OJ No. L192 (1975).

Council Aid to Shipbuilding Directive No. 78/338/EEC OJ No. L98. (1978).
Council Decision No. 78/774/EEC OJ No. L258 (1978) and Council Decision No. 79/4/EEC OJ No L5 (1979).
Council Aid to Shipbuilding Directive No. 81/363/EEC OJ No. L137 (April 1981) as extracted by Council Directive No. 82/880/EEC OJ No. L3371 (21 December 1982).
Proposals for a Community Maritime Policy EC Doc. No 5635/85 (March 1985).
First phase of EEC Common Shipping Policy, Comprising Regulations 4055/86, 4056/86, 4057/86 and 4058/86 (July 1986).
Council Aid to Shipbuilding Directive No. 87/167/EC (January 1987).
Proposals for Second Stage of the Common Maritime Transport Policy (EC Docs 8367/89 and 8368/89 (August 1989).
Council Aid to Shipbuilding Directive No. 90/684/EC OJ L374 (December 1990).

Parliamentary Debates (Official Report, Hansard)

Fifth Series (5 Ser), House of Commons, Vols 1 to 1,000, February 1909 to March 1981.
Sixth Series (6 Ser), House of Commons from March 1981 and since.

Secondary Sources

Place of publication is London unless otherwise stated.

Abell, W.A., *The Shipwright's Trade* (Cambridge, 1948).
Aberconway, Lord, *The Basic Industries of Great Britain* (1927).
Albu, A., 'Causes of the Decline in British Merchant Shipbuilding and Marine Engineering', *Omega*, 4 (1976).
Albu, A., 'Merchant Shipbuilding and Marine Engineering' in K. Pavitt (ed.), *Technical Innovation and British Economic Performance* (1984).
Aldcroft, D.H., 'The Decontrol of British Shipping and Railways After the First World War', *Journal of Transport History*, 5 (1961)
Aldcroft, D.H., 'The Eclipse of British Coastal Shipping, 1913–1921', *Journal of Transport History*, 6 (1963).
Aldcroft, D.H., *The Interwar Economy: Britain, 1919–1939* (1970).
Aldcroft, D.H., *Studies in British Transport History, 1870–1970* (Newton Abbot, 1974).
Aldcroft, D.H., *British Transport since 1914: An Economic History* (Newton Abbott, 1975).
Alexander, K.J.W. and Jenkins, C.L., *Fairfield's: A Study in Institutional Change* (1970).
Armitage, A., 'Shipbuilding at Belfast: Workman, Clark and Company, 1880–1935' in L.R. Fischer (ed.), *From Wheel House to Counting House: Essays in Maritime Business History in Honour of Professor Peter Neville Davis* (St John's, 1992).
Armstrong, J., 'Climax and Climacteric: The British Coastal Trade, 1870–1930' in A.G. Jamieson and D.J. Starkey (eds), *Exploiting the Sea* (Exeter, 1998).
Armstrong, P., Glynn, A. and Harrison, J., *Capitalism Since 1945* (Oxford, 1991).
Arnold, A.J., '"Of Not Much Significance": The Inclusion of Naval and Foreign

Business in the UK Shipbuilding Output Series, 1825–1914', *International Journal of Maritime History*, 11 (1999).
Arnold, A.J., *Iron Shipbuilding on the Thames, 1832–1915* (Aldershot, 2000).
Arnold, G., *Britain's Oil* (1978).
Ashworth, W., *Contracts and Finance* (1953).
Atkinson, R., *The Development and Decline of British Shipbuilding* (1999).
Ayre, Sir W., *A Shipbuilder's Yesterdays* (Aberdour, 1968).
Barnaby, K.C., *One Hundred Years of Specialised Shipbuilding and Engineering: The History of John I. Thornycroft and Company Ltd.* (Southampton, 1964).
Barnett, C., *The Audit of War* (1986).
Barnett, C., *Engage the Enemy More Closely: The Royal Navy in the Second World War* (1991).
Behrens, C.B.A., *Merchant Shipping and the Demands of War* (1955).
Bell, C.M., *The Royal Navy, Seapower and Strategy Between the Wars* (2000).
Benn, T., *Office Without Power: Diaries 1968–1972* (1988).
Benwell Community Development Project, *The Making of a Ruling Class* (Newcastle-upon-Tyne, 1978).
Benwell Community Development Project, *The Costs of Industrial Change* (no date).
Blackaby, F., (ed.), *British Economic Policy, 1960–1974* (Cambridge, 1978).
Blake, G., *Down to the Sea* (1937).
Blake, G., *British Ships and Shipbuilders* (1946).
Boyce, G.H., *Information, Mediation and Institutional Development: The Rise of Large-Scale Enterprise in British Shipping, 1870–1919* (Manchester, 1995).
Boyce, G.H., 'Union Steamship Company of New Zealand and the Adoption of Oil Propulsion: Learning-by-Using Effects', *Journal of Transport History*, 18 (September 1997).
Bredima-Savopoulou, A. and Tzoannos, J., *The Common Shipping Policy of the EC* (Amsterdam, 1990).
Broadway, F., *Upper Clyde Shipbuilders: A Study of Government Intervention in Industry* (1976).
Broeze, F., *Island Nation: A History of Australians and the Sea* (St Leonards, NSW, 1998).
Broeze, F., 'Containerization and the Globalization of Liner Shipping' in D.J. Starkey and G. Harlaftis (eds), *Global Markets: The Internationalization of the Sea Transport Industry since 1850* (St John's, 1998).
Brown, D.K., *A Century of Naval Construction* (1983).
Buchan, A., *The Right to Work: The Story of the Upper Clyde Confrontation* (1972).
Burn, D.L., *The Economic History of Steel Making, 1867–1939* (Cambridge, 1940).
Burn, D.L., *The Steel Industry, 1939–1959* (Cambridge, 1961).
Burrell, D., *Furness Withy, 1891–1991* (Kendal, 1992).
Burton, A., *The Rise and Fall of British Shipbuilding* (1994).
Buxton, I.L., 'The Development of the Merchant Ship, 1880–1990', *Mariner's Mirror*, 79 (1993).
Buxton, I.L., *Warship Building and Repair during the Second World War* (Glasgow, 1998).
Buxton, N.K., 'The Scottish Shipbuilding Industry between the Wars: A Comparative Study', *Business History*, 10 (1968).
Buxton, N.K., 'The Scottish Economy, 1945–79: Performance, Structure and

Problems' in R. Saville (ed.), *The Economic Development of Modern Scotland, 1950–1980* (Edinburgh, 1985).

Cairncross, A.K., *The British Economy since 1945* (Oxford, 1992).

Cairncross, A.K. and Parkinson, J.R., 'The Shipbuilding Industry' in D. Burn (ed.), *The Structure of British Industry*, Vol, 2 (Cambridge, 1958).

Calfruny, A.W., *Ruling the Waves: The Political Economy of International Shipping* (Berkley, 1987).

Cammell Laird, *Builders of Great Ships: The History of Cammell Laird and Company Ltd* (Birkenhead, 1959).

Campbell, R.H., *The Rise and Fall of Scottish Industry, 1707–1939* (Edinburgh, 1980).

Campbell, R.H., 'Costs and Contracts: Lessons from Clyde Shipbuilding Between the Wars' in A. Slaven and D.H. Aldcroft (eds), *Business, Banking and Urban History: Essays in Honour of S.G. Checkland* (Edinburgh, 1982).

Carvel, J.L., *Stephen of Linthouse 1750–1950* (Glasgow, 1950).

Cayzer, N., 'The Challenge to British Shipping', *Institute of Transport Journal*, 30 (1963).

Chalmers, M., *Paying for Defence: Military Spending and British Decline* (1985).

Chida, T. and Davies, P.N., *The Japanese Shipping and Shipbuilding Industries* (Berekley, 1987).

Chrzanowski, I., Krzyanowski, M. and Krzysztof, L., *Shipping Economics and Policy: A Socialist View* (1983).

Clarke, A., *Sunderland Shipyards* (Seaham, 1998).

Clarke, J.F., *Building Ships on the North East Coast*, Vol. 2 (Whitley Bay, 1997).

Clarke, J.F., *Power on Land and Sea* (Newcastle-upon-Tyne, no date).

Clay, Sir H., *Lord Norman* (1957).

Clegg, H.A., *The Changing System of Industrial Relations in Great Britain* (Oxford, 1979).

Coates, D. and Hillard, J. (eds), *The Economic Decline of Modern Britain* (Hemel Hempstead, 1986).

Cooper, B. and Gaskell, T.F., *North Sea Oil: The Great Gamble* (1966).

Coopey, R., Fielding, S. and Tiratsoo, N. (eds), *The Wilson Governments, 1964–1970* (1993).

Corlett, E., *The Ship: The Revolution in Merchant Shipping* (1981).

Corley, T.A.B., *A History of the Burmah Oil Company: Vol. 11, 1924–1966* (1988).

Coster, G., *Corsairville: The Lost Domain of the Flying Boat* (2000).

Craig, R., *The Ship: Steam Tramps and Cargo Liners* (1980).

Crossman, R.H.S., *Diaries of a Cabinet Minister, Vol. 1, Minister of Housing, 1964–66* (1976).

Davies, M., *Belief In the Sea* (1992).

Davies, P.N., *The Trade Makers: Elder Dempster in West Africa 1852–1972* (1973).

Davies, P.N., 'The Role of National Bulk Carriers in the Advance of Shipbuilding Technology in Post-War Japan', *International Journal of Maritime History*, Vol 4 (June 1992).

Davies, P.N. and Bourn, A.M., 'Lord Kylsant and the Royal Mail', *Business History*, 14 (1972).

Denholm, I., 'The Decline of British Shipping: A Personal View', *Maritime Policy and Management*, 17 (1991).

Denny Brothers, *Denny, Dumbarton 1844–1932* (Dumbarton, 1932).

Dougan, D., *The History of North-East Shipbuilding* (1968).
Dougan, D., *The Shipwrights* (Newcastle-Upon-Tyne, 1975).
Drury, C. and Stokes, P., *Ship Finance: The Credit Crisis* (1983).
Elbaum, B. and Lazonick, W. (eds),*The Decline of the British Economy* (Oxford, 1986).
Ennis, F. and Roberts, I., 'The Time of their Lives? Female Workers in North East Shipbuilding, 1939–1945' in A. Potts (ed.), *Shipbuilders and Engineers: Essays in Labour History in the Shipbuilding and Engineering Industries in the North East of England* (Durham, 1987).
Evans, H., *Vickers: Against the Odds* (1978).
Farrington, T., 'The UK Shipping Industry: A History and Assessment', *The Business Economist*, 27 (1996).
Fassett, H.G. (ed.), *The Shipbuilding Business in the United States of America*, 2 Vols (New York, 1948).
Fayle, C.E., *The War and the Shipping Industry* (Oxford, 1927).
Fforde, J., *The Bank of England and Public Policy, 1941–1958* (Cambridge, 1992).
Flanders, A., *The Fawley Productivity Agreements: A Case Study of Management and Collective Bargaining* (1964).
Flanders, A., *Management and Unions* (1970).
Fletcher, M.E., 'From Coal to Oil in British Shipping', *Journal of Transport History*, 3 (1975).
Foot, P., *The Politics of Harold Wilson* (Harmondsworth, 1968).
Foster, J. and Woolfson, C., *The Politics of the UCS Work-In* (1986).
Gibson, A. and Donovan, A.,*The Abandoned Ocean: A History of United States Maritime Policy* (Columbia, 2000).
Gilmour, I., *Dancing with Dogma: Britain Under Thatcherism* (1992).
Gilmour, I and Garnett, M., *Whatever Happened to the Tories* (1997).
Glyn, A. and Sutcliffe, R., *British Capitalism, Workers and the Profit Squeeze* (Harmondsworth, 1972).
Goodey, C., *The First Hundred Years: The Story of Richards Shipbuilders* (Lowestoft, 1976).
Gordon, G.A.H., *British Seapower and Procurement between the Wars: A Reappraisal of Rearmament* (1988).
Gorst, A. and Johnman, L., 'Naval Procurement and British Shipbuilding, 1945–1964' in A.G. Jamieson and D.J. Starkey (eds), *Exploiting the Sea* (Exeter, 1998).
Goss, R.O., *Advances in Maritime Economics* (Cambridge, 1982).
Goss, R.O., 'The Decline of British Shipping: A Case for Action? And a Comment', *Maritime Policy and Management*, 20 (1993).
Grangemouth Dockyard, *Grangemouth Dockyard Co. Ltd* (Grangemouth, 1951).
Grant, A., *Steel and Ships: The History of John Brown's* (1950).
Green, E. and Moss, M., *A Business of National Importance: The Royal Mail Shipping Group 1902–1937* (1982).
Green, F. (ed.), *The Restructuring of the UK Economy* (1989).
Greenhill, B., *The Ship: The Life and Death of the Merchant Sailing Ship, 1815–1965* (1980).
Greenway, A. (ed.), *Conway's History of the Ship—The Golden Age of Shipping: The Classic Merchant Ship, 1900–1960* (1994).

Gregory, R.H., 'Court Line, Mr Benn and the Ombudsman', *Parliamentary Affairs*, 30 (1977).
Griffiths, D., 'British Shipping and the Diesel Engine: The Early Years', *Mariner's Mirror*, 81 (1995).
Grove, E.J., *Vanguard to Trident: British Naval Policy Since World War Two* (Annapolis, 1987).
Grove, E.J. (ed.), *The Defeat of the Enemy Attack on Shipping 1939–45* (Aldershot, 1998).
Hamilton, A., *North Sea Impact: Offshore Oil and the British Economy* (1978).
Hancock, W.K. and Gowing, M.M., *British War Economy* (1949).
Hannah, L., *The Rise of the Corporate Economy* (1983).
Harlaftis, G., *A History of Greek-Owned Shipping* (1996).
Harris, L., *A Two Hundred Year History of Appledore Shipyards* (Combe Martin, 1992).
Harvie, C., *Fool's Gold: The Story of North Sea Oil* (Harmondsworth, 1994).
Henning, G. and Trace, K., 'Britain and the Motor Ship: A Case of the Delayed Adoption of a New Technology?', *Journal of Economic History*, 35 (1975).
Herron, F., *Labour Market in Crisis: Redundancy at Upper Clyde Shipbuilders* (1975).
Hilditch, P., 'The Decline of British Shipbuilding since the Second World War' in S. Fisher (ed.), *Lisbon as a Port Town, The British Seaman, and other Maritime Themes* (Exeter, 1988).
Hilditch, P.J. and Reid, A.J., 'Trade Unions and Labour Productivity: The British Shipbuilding Industry, 1870–1950', *DAE Working Paper No. 8907* (Cambridge, 1989).
Hill, J.C.G., *Shipshape and Bristol Fashion, Charles Hill & Sons Ltd* (Bristol, 1951).
HMSO, *Build the Ships: The Official Story of the Shipyards in Wartime* (1947).
HMSO, *Statistical Digest of the War* (1951).
Hogwood, B., *Government and Shipbuilding: The Politics of Industrial Change* (Farnborough, 1979).
Hollett, D., *Men of Iron: The Story of Cammell Laird Shipbuilders, 1828–1991* (Birkenhead, 1992).
Holms, C.A., *Practical Shipbuilding*, 2 Vols (1916).
Hope, R., *A New History of British Shipping* (1990).
Hornby, W., *Factories and Plant* (1958).
Hume, J.R. and Moss, M.S., *Beardmore: The History of a Scottish Historical Giant* (1979).
Hyde, F.E., *Shipping Enterprise and Management 1830–1939: Harrisons of Liverpool* (Liverpool, 1967).
Hyde, F.E., *Cunard and the North Atlantic, 1840–1973* (1975).
Inman, P., *Labour in the Munitions Industries* (1957).
Isserlis, L., 'Tramp Shipping, Cargoes and Freights', *Journal of the Royal Statistical Society*, 101 (1938).
Jamieson, A.G., 'British OSV Companies in the North Sea, 1964–1997', *Maritime Policy and Management*, 25 (1988).
Jamieson, A.G., 'Facing the Rising Tide: British Attitudes to Asian National Shipping Lines, 1959–1964', *International Journal of Maritime History*, 7 (1995).
Jamieson, A.G., *Ebb Tide* (Exeter, 2003, forthcoming).
Jenkin, M., *British Industry and the North Sea: State Intervention in a Developing Industrial Sector* (1981).

Jenkins, J.G. and Jenkins, D., *Cardiff Shipowners* (Cardiff, 1986).
Johnman, L., 'The Labour Party and Industrial Policy, 1940–1945' in N. Tiratsoo (ed.), *The Attlee Years* (1991).
Johnman, L., 'The Shipbuilding Industry' in H. Mercer, N. Rollings and J. Tomlinson (eds), *Labour Governments and Private Industry: The Experience of 1945–51* (Edinburgh, 1992).
Johnman, L., 'The Privatisation of British Shipbuilders', *International Journal of Maritime History*, 8 (December 1996).
Johnman, L., 'Old Attitudes and New Technology: British Shipbuilding, 1945–1965' in P.C. van Royen, L.R. Fischer and D.M. Williams (eds), *Fruitta di Mare: Evolution and Revolution in the 19th and 20th Centuries* (Amsterdam, 1998).
Johnman, L., 'Internationalization and the Collapse of British Shipbuilding, 1945–1965' in D.J. Starkey and G. Harlaftis (eds), *Global Markets: The Internationalization of the Sea Transport Industries Since 1850* (St John's, 1998).
Johnman, L., *British Shipbuilding and the State* (Glasgow, 1998).
Johnman, L. Johnston, I. and Mackenzie, I., *Down the River* (Glendaruel, 2001).
Johnman, L. and Murphy, H., 'The Norwegian Market for British Shipbuilding, 1945–1967', *Scandinavian Economic History Review*, 46 (1998).
Johnman, L. and Murphy, H., 'No Light at the End of the Dock: The Long Rise and Short Life of the Firth of Clyde Dry Dock Company', *International Journal of Maritime History*, 10 (December 1998).
Johnman, L. and Murphy, H., 'A Very British Institution! A Study in Under-Capitalisation: The Role of the Ship Mortgage Finance Company in Post-Delivery Credit Financing in British Shipbuilding, 1951–1967', *Financial History Review*, 6 (October 1999).
Johnman, L. and Murphy, H., 'The Rationalisation of Warship Building in the United Kingdom, 1945–2000', *Journal of Strategic Studies*, 24 (2001).
Johnman, L. and Murphy, H., 'A Triumph of Failure: The British Shipbuilding Industry and the Offshore Structures Market, 1960–1990: A Case Study of Scott Lithgow Ltd', *International Journal of Maritime History* (forthcoming, 2002).
Johnson, C., *The Economy Under Mrs Thatcher 1979–1990* (Harmondsworth, 1991).
Johnston, I., *Beardmore Built: The Rise and Fall of a Clydeside Shipyard* (Clydebank, 1993).
Johnston, I., *Ships for a Nation, 1847–1971: John Brown & Company, Clydebank* (West Dumbartonshire, 2000).
Jones, A., *Oil: The Missed Opportunity* (1981).
Jones, L., *Shipbuilding in Britain: Mainly Between the Two World Wars* (Cardiff, 1957).
Jones, S., *Two Centuries of Overseas Trading: The Origins and Growth of the Inchcape Group* (1986).
Jones, S., 'The P & O in War and Slump, 1914–1932: The Chairmanship of Lord Inchcape' in S. Fisher (ed.), *Innovation in Shipping and Trade* (Exeter, 1989).
Jung, I., *The Marine Turbine*, 3 Vols (1986).
Kaldor, N., *Causes of Slow Growth of the United Kingdom* (Cambridge, 1966).
Kennedy G.C., 'Great Britain's Maritime Strength and the British Merchant Marine, 1922–1935', *Mariner's Mirror*, 80 (1994).
Kennedy G.C. (ed.), *The Merchant Marine in International Affairs, 1850–1950* (2000).
Kennedy, P., *The Rise and Fall of British Naval Mastery* (1976).
Kirby, M.W., *The British Coalmining Industry 1870–1946* (1977).

Kirby, M.W., *The Decline of British Economic Power Since 1870* (1981).
Kirby, M.W., 'Institutional Rigidities and Economic Decline : Reflections on the British Experience', *Economic History Review*, 65 (November 1992).
Kirkaldy, A.W., *British Shipping* (1914).
Kirkwood, D., *My Life of Revolt* (1935).
Kuuse, J. and Slaven, A. (eds), *Scottish and Scandinavian Shipbuilding: Development Problems in Historical Perspective* (Glasgow, 1980).
Lambert, N., *Sir John Fisher's Naval Revolution* (Columbia, 1999).
Lane, F.C., *Ships for Victory: A History of Shipbuilding under the US Maritime Commission in World War II* (Baltimore, 1951).
Lechmann, J., *A Century of Burmeister & Wain* (Copenhagen, 1948).
Lingwood, J., *SD14—The Great British Shipbuilding Success Story* (1976).
Lipscomb, F.W., *The British Submarine* (1975).
Lords Commissioners of the Admiralty, *Electric Welding in Shipbuilding* (1943).
Lorenz, E.H., 'Two Patterns of Development: The Labour Process in the British and French Shipbuilding Industries, 1880–1930', *Journal of European Economic History*, 13 (1984).
Lorenz, E.H., *Economic Decline In Britain: The Shipbuilding Industry, 1890–1970* (Oxford, 1991).
Lorenz, E.H. and Wilkinson, F., 'The Shipbuilding Industry, 1880–1965' in B. Elbaum and W. Lazonick (eds), *The Decline of the British Economy* (Oxford, 1986).
Lovell, J., 'Employers and Craft Unionism: A Programme of Action for British Shipbuilding, 1902–5', *Business History*, 34 (1992).
Lucas, A.F., *Industrial Reconstruction and the Control of Competition* (1937).
Lyon, D.J., *The Denny List* (1975).
Lyon, D.J., *The Ship: Steam, Steel and Torpedoes* (1980).
Lyon, D.J., *The First Destroyers* (1996).
Lyon, H., 'The Admiralty and Private Industry' in B. Ranft (ed.), *Technical Change and British Naval Policy, 1860–1939* (1977).
MacKay, D.I. and MacKay, G.A., *The Political Economy of North Sea Oil* (1975).
MacDonald, I. and Tabner, L., *Smith's Dock* (Seaworks, 1986).
McCord, N. (ed.), *Essays in Tyneside Labour History* (Newcastle, 1977).
McGill, J., *Crisis on the Clyde: The Story of Upper Clyde Shipbuilders* (1973).
McGoldrick, J., 'Crisis and the Division of Labour in the Shipbuilding Industry Since the War', *British Journal of Industrial Relations*, 21 (1983).
McKeown, D., 'Welding—The Quiet Revolution' in A. Slaven and F. Walker (eds), *European Shipbuilding: One Hundred Years of Change* (1983).
McKinlay, A., 'The 1937 Apprentices Strike: Challenge from an Unexpected Quarter', *Journal of Scottish Labour History Society*, 20 (1985).
McKinlay, A., 'From Industrial Serf to Wage Labourer: The 1937 Apprentice Revolt in Britain', *International Review of Social History*, 31 (1986).
McKinlay, A., 'The Interwar Depression and the Effort Bargain: Shipyard Riveters and the Workers Foreman, 1919–1939', *Scottish Economic and Social History*, 19 (1989).
McKinstry, S., 'Transforming John Brown's Shipyard: The Drilling Rig and Offshore Fabrication Business of Marathon and UIE, 1972–1977', *Scottish Economic and Social History*, 18 (1998).

McMurray, H.C., 'The Records of the Shipbuilders and National Repairers Association', *Business Archives*, 45, (1979).
Maber, J.M., *The Ship: Channel Packets and Ocean Liners, 1850–1970* (1980).
Maddison, A., *Phases of Capitalist Development* (Oxford, 1982).
Maddison, A., *Dynamic Forces in Capitalist Development* (Oxford, 1991).
Maiolo, J.A., *The Royal Navy and Nazi Germany, 1933–39* (1998).
Marder, A.J., *From the Dreadnought to Scapa Flow: The Royal Navy in the Fisher Era, 1904–1919*, 5 Vols (1961–70).
Maynard, G., *The Economy Under Mrs. Thatcher* (Oxford, 1988).
Methayas, B.N., *The Economics of Tramp Shipping* (1971).
Michie, J. (ed.), *The Economic Legacy 1979–1992* (1992).
Middlemass, K., *Power, Competition and the State*, 3 Vols (1986, 1990 and 1991).
Middlemass, N., *British Shipbuilding Yards*, 3 Vols (South Shields 1993, 1994 and 1995).
Milward, A.S. , *The Economic Effects of the World Wars on Britain* (1970).
Mitchell, B.R. and Deane, P., *Abstract of British Historical Statistics* (Cambridge, 1962).
Money, Sir L., 'The Nationalisation of Shipping', *English Review*, 28 (1919).
Morgan, D.J., 'Boom and Slump: Shipowning in Cardiff, 1919–1921', *Maritime Wales*, 12 (1989).
Mortimer, J.E., *History of the Boilermakers' Society*, 3 Vols (1973, 1981 and 1984).
Moss, M. and Hume, J.R., *Shipbuilders to the World: 125 Years of Harland and Wolff, Belfast 1861–1986* (Belfast, 1986).
Murphy, H., '"From the Crinoline to the Boilersuit": Women Workers in British Shipbuilding During the Second World War', *Contemporary British History*, 13 (1999).
Murphy, H., 'Lost Opportunities: Women in Britain's Private Wartime Shipyards', *American Neptune*, 60 (2000).
Murphy, H., 'Scotts of Greenock and Naval Procurement 1945–1977', *Mariner's Mirror*, 87 (2001).
Newman, B., *Plate and Section Working Machinery in British Shipbuilding, 1850–1945* (Glasgow, 1983).
Nicolas, R., *Changing Tide: The Final Years of Wear Shipbuilding* (Sunderland, 1990).
Noel-Baker, P., *The Private Manufacture of Armaments* (1937).
North Tyneside Trades Council, *Shipbuilding: The Cost of Redundancy* (Newcastle, 1979).
OECD, *The Situation of the Shipbuilding Industry* (Paris, 1965).
O'Hara, G.C., *Ironfighters, Outfitters and Bowler Hatters* (Prestwick, 1997).
Ollson, K, 'Big Business in Sweden: The Golden Age of the Great Swedish Shipyards, 1945–1974', *Scandinavian Economic History Review*, 43 (1993).
Orbell, J., *From Cape to Cape: The History of Lyle Shipping* (Edinburgh, 1978).
Palmer, S., 'The British Coal Export Trade, 1850–1913' in D. Alexander and R. Ommer (eds), *Volumes not Values: Canadian Ships and World Trades* (St John's, 1979).
Parker, G., *At the Sharp End! A Shipbuilding Autobiography* (Glasgow 1992).
Parker, H.M.D., *Manpower: A Study of War-Time Policy and Administration Policy* (1957).
Parkinson, J.R., 'Trends in the Output and Export of Merchant Ships', *Scottish Journal of Political Economy*, 3 (1956).

Parkinson, J.R., *The Economics of Shipbuilding in the United Kingdom* (Cambridge, 1960).
Parkinson, J.R., 'The Financial Prospects of Shipbuilding After Geddes', *Journal of Industrial Economics*, 17 (1968).
Parkinson, J.R., 'Shipbuilding' in N.K. Buxton and D.H. Aldcroft (eds), *British Industry Between the Wars* (1979).
Paulden, S. and Hawkins, B., *Whatever Happened at Fairfield's?* (1969).
Payne, P.L., *Colvilles and the Scottish Steel Industry* (Oxford, 1979).
Payton-Smith, D.J., *Oil* (1971).
Peden, G.C., *British Rearmament and the Treasury, 1932–1939* (Edinburgh, 1979).
Peebles, H., *Warshipbuilding on the Clyde* (Edinburgh, 1987).
Pickard, T., *We Build Ships* (1989).
Political and Economic Planning, 'The British Shipping Industry', *Planning*, 25 (1959).
Pollard, S., 'British and World Shipbuilding, 1890–1914: A Study in Comparative Costs', *Journal of Economic History*, 17 (1951).
Pollard, S., 'Laissez-Faire and Shipbuilding', *Economic History Review*, 5 (1952).
Pollard, S. and Robertson, P.L., *The British Shipbuilding Industry, 1890–1914* (Cambridge, Mass., 1979).
Porter, M. (ed.), *Cases of Competitive Strategy* (New York, 1983).
Postan, M.M., *British War Production* (1952).
Preston, A., *The Ship: Dreadnought to Nuclear Submarine* (1980).
Pugh, P., *The Cost of Seapower* (1986).
Purvis, M.K., 'The Postwar Royal Navy Frigate and Guided Missile Destroyer Design, 1945–1969', *Transactions of the Royal Institute of Naval Architects* (1974).
Rabson, S. and O'Donoghue, K., *P&O: A Fleet History* (Kendal, 1988).
Ranft, B. (ed.), *Technical Change and British Naval Policy, 1860–1939* (1977).
Rapping, L., 'Learning and World War II Production Functions', *Review of Economics and Statistics*, 48 (1965).
Ratcliffe, M., *Liquid Gold Ships: A History of the Tanker, 1859–1984* (Colchester, 1985).
Readheads, J., *Readheads, 1865–1965* (South Shields, 1965).
Reid, J.M., *James Lithgow: Master of Work* (1964).
Riddle, I., *Shipbuilding Finance* (Cardiff, 1983).
Ritchie, L.A., *The Shipbuilding Industry: A Guide to Historical Records* (Manchester, 1992).
Roberts, D., *Life at Lairds* (Bebington, 1992).
Roberts, I., *Craft, Class and Control* (Edinburgh, 1993).
Robertson, A.J., 'Backward British Businessmen and the Motor Ship, 1918–1939: The Critique Reviewed', *Journal of Transport History*, 9 (1988).
Robertson, D.J., 'Labour Turnover in the Clyde Shipbuilding Industry', *Scottish Journal of Political Economy*, 1 (1954).
Robertson, P.L., 'Shipping and Shipbuilding: The Case of William Denny and Brothers', *Business History*, 16 (1964).
Robertson, P.L., 'Technical Education in the British Shipbuilding and Marine Engineering Industries, 1863–1914', *Economic History Review*, 27 (1974).
Robinson, C. and Hann, D., 'North Sea Oil and Gas' in P. Johnson (ed.), *The Structure of British Industry* (1988).
Roskill, S., *The War at Sea*, 3 Vols (1954–61).

Roth, A., *Heath and the Heathmen* (1972).
Safford, I., 'Anglo-American Maritime Relations During the Two World Wars: A Comparative Analysis', *American Neptune*, 41 (1981).
Sampson, A., *The Seven Sisters* (1988).
Sankey, R., *Maritime Heritage: Barrow and Morecambe Bay* (Kettering, 1986).
Savage, C.I., *Inland Transport* (1957).
Saville, R., *The Bank of Scotland: A History, 1695–1995* (Edinburgh, 1996).
Sayers, R.S., *The Bank of England, 1891–1944* (Cambridge, 1976).
Scott, J.D., *Vickers: A History* (1962).
Scotts of Greenock, *Two Hundred and Fifty Years of Shipbuilding by the Scotts of Greenock* (Greenock, 4th edn, 1961).
Shay, R.B., *British Rearmament in the Thirties: Politics and Profits* (1977).
Shields, J., *Clyde Built* (Glasgow, 1948).
Shonfield, A., *British Economic Policy Since the War* (Harmondsworth, 1958).
Short, J., 'Defence Spending in the UK Regions', *Regional Studies*, 15, (1981).
Slaven, A., *The Development of the West of Scotland: 1750–1960* (1975).
Slaven, A., 'A Shipyard in Depression: John Brown's of Clydebank, 1919–38', *Business History*, 19 (1977).
Slaven, A., 'Management and Shipbuilding, 1890–1938: Structure and Strategy in the Shipbuilding Firm on the Clyde' in A. Slaven and D.H. Aldcroft, *Business, Banking and Urban History: Essays in Honour of S.G. Checkland* (Edinburgh, 1982).
Slaven, A., 'Self-Liquidation: The National Shipbuilding Security Ltd and British Shipbuilders in the 1930s' in S. Palmer and G. Williams (eds), *Charted and Uncharted Waters: Proceedings of a Conference on the Study of Maritime History* (1982).
Slaven, A., 'British Shipbuilders: Market Trends and Order Book Patterns Between the Wars', *Journal of Transport History*, 3 (1982).
Slaven, A., 'Marketing Opportunities and Marketing Practices: The Eclipse of British Shipbuilding, 1957–1976' in L.R. Fischer (ed.), *From Wheel House to Counting House: Essays in Maritime Business History in Honour of Professor Peter Neville Davies* (St John's, 1992).
Slaven, A. and Walker, F. (eds), *The Fourth International Shipbuilding and Ocean Engineering Conference* (Gothenburg, 1986).
Smith, J.W. and Holden, T.S., *Where Ships Are Born, Sunderland, 1346–1946* (Sunderland, 1946).
Somner, G., *Ben Line: Fleet and Short History* (Kendal, 1980).
Somner, G., *From 70 North to 70 South: A History of the Christian Salvesen Fleet* (Edinburgh, 1984).
Speed, K., *Sea Change* (Bath, 1982).
Stokes, P., *Ship Finance: Credit Expansion and the Boom–Bust Cycle* (1992).
Stone, I., *Shipbuilding on Wearside* (Sunderland, 1988).
Stopford, M., *Maritime Economics* (1988).
Stopford, R.M. and Barton, J.R., 'Economic Problems of Shipbuilding and the State', *Maritime Policy and Management*, 13 (1986).
Strath, B., *The Politics of De-industrialisation: The Contraction of the West European Shipbuilding Industry* (1987).
Sturmey, S.G., *British Shipping and World Competition* (1962).
Sturmey, S.G., *Shipping Economics* (1975).

Summerfield, P., *Women Workers in the Second World War: Production and Patriarchy in Conflict* (1984).
Sumida, J.T., *In Defence of Naval Supremacy* (1989).
Supple, B.E., *The History of the British Coal Industry: Vol. 4: 1913–1946, The Political Economy of Decline* (Oxford, 1987).
Swann, D., *Competition and Industrial Policy in the European Community* (1983).
Tenold, S. and Nordvik, H.W., 'Coping with the International Shipping Crisis of the 1990s: A Study of Management Responses in Norwegian Oil Tanker Companies', *International Journal of Maritime History*, 8, 2 (1996).
Thomas, D., 'Shipbuilding—Demand Linkage and Industrial Decline' in K. Williams *et al.*, *Why are the British Bad at Manufacturing?* (1983).
Thompson, M. *et al.*, *Cook, Welton and Gemmell: Shipbuilders of Hull and Beverley, 1883–1963* (Beverley, 1999).
Thompson, W. and Hart, F., *The UCS Work-In* (1972).
Thornton, R.H., *British Shipping* (Cambridge, 1959).
Tipper, C., *The Brittle Fracture Story* (Cambridge, 1962).
Todd, D., 'Regional Variations in Naval Construction: The British Experience, 1895–1996', *Regional Studies*, 15, 2 (1981).
Todd, D., *The World Shipbuilding Industry* (New York, 1986).
Todd, D., *Industrial Dislocation: The Case of Global Shipbuilding* (1991).
Tolliday, S., *Business, Banking and Politics: The British Steel Industry in the Interwar Years* (Cambridge, Mass., 1987).
Tomlinson, J., *Public Policy and the Economy Since 1900* (Oxford, 1990).
Tomlinson, J., '"Inventing Decline": The Falling Behind of the British Economy in the Post-war Years', *Economic History Review*, 69 (1996).
Trebilcock, C., *The Vickers Brothers, Armaments and Enterprise* (1977).
Vamplew, W., *Salvesen of Leith* (Edinburgh, 1974).
Van Der Wee, H., *Prosperity and Upheaval: The World Economy, 1945–1980* (Harmondsworth, 1987).
Ville, S. (ed), *Shipbuilding in the United Kingdom in the Nineteenth Century: A Regional Approach* (St John's, 1993).
Volk, B., *The Shipbuilding Cycle—A Phenomenon Explained* (Bremen, 1994).
Walker, F.M., *Song of the Clyde: A History of Clyde Shipbuilding* (1984).
Walker, F.M. and Slaven, A. (eds), *European Shipbuilding: One Hundred Years of Change* (1984).
Warren, K., *Armstrongs of Elswick* (1990).
Warren, K., *Steel, Ships and Men* (Liverpool, 1998).
White, J. Samuel and Company, *From Smack to Frigate: From Cutter to Destroyer* (Cowes, 1928).
Whitehurst, C.H., *The US Shipbuilding Industry: Past, Present and Future* (Annapolis, 1986).
Wilkinson, E., *The Town That Was Murdered: A History of Jarrow* (1939).
Williams, D.L., *Maritime Heritage: White's of Cowes* (Peterborough, 1993).
Withington, J., *Shutdown: The Anatomy of a Shipyard Closure* (1989).
Workman Clark, *Shipbuilding at Belfast 1880–1933: History of Workman Clark (1928) Ltd* (Belfast, 1934).
Wren, C., 'Gross Expenditure on UK Industrial Assistance: A Research Note',

Scottish Journal of Political Economy, 43 (1996).
Yamshita, Y., 'Responding to the Global Market in Boom and Recession: Japanese Shipping and Shipbuilding Industries, 1945–1980' in S.P. Ville and D.M. Williams (eds), *Management, Finance and Industrial Relations in the Maritime Industries: Essays in International Maritime and Business History* (St John's, 1994).
Yarrow and Company, *Yarrow and Co. Ltd. 1865–1977* (Glasgow, 1977).
Yarrow, Lady, *Alfred Yarrow: His Life and Work* (1923).
Yergin, D., *The Prize: The Epic Quest for Oil, Money and Power* (1991).

Index

Printed and bound by CPI Group (UK) Ltd, Croydon, CR0 4YY

27/10/2024

14580409-0004